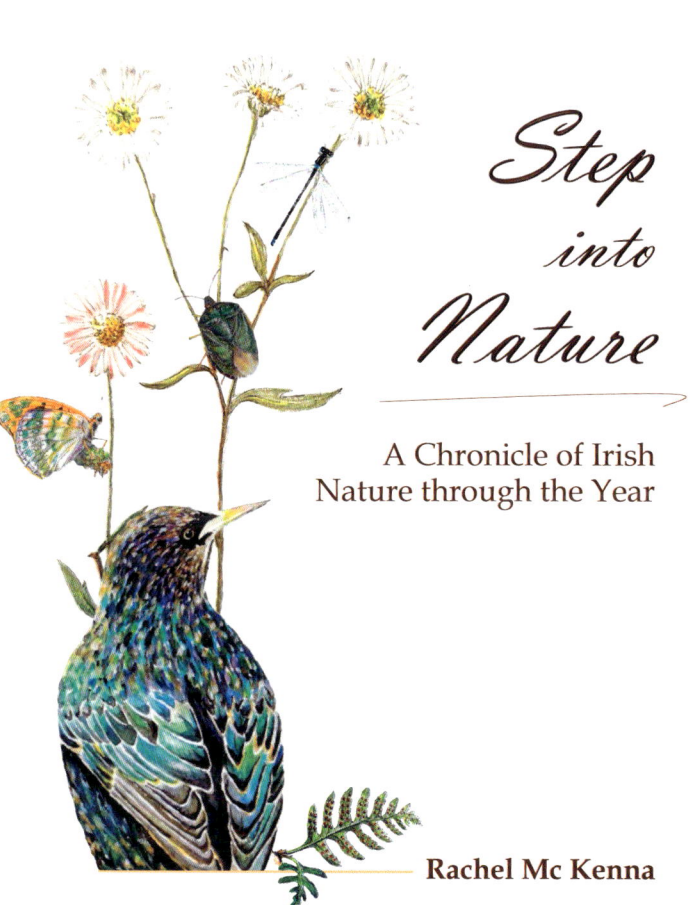

Step into Nature

A Chronicle of Irish Nature through the Year

Rachel Mc Kenna

Wild Carrot, *Daucus carota*

Step into Nature

© 2024 Rachel Mc Kenna
Reprinted December 2025

Front cover images: Polypody Fern, Starling, Silver washed Fritillary, Green Shieldbug and Mexican Fleabane

Back cover images: Long-tailed Tit and hoverfly, *Sphaerophoria scripta*

Text, illustrations, photographs and graphic design by Rachel Mc Kenna. Permission to use or reprint any of this publication must be obtained from the author.

Printed in Ireland for Dimma Print Services, Roscrea, Co. Tipperary, 2025.

Distributed by Mercier Press.

ISBN 9781781179529

This book is supported by the Heritage Council, and is an action of the Offaly Heritage Plan 2023-2027.

Foreword

There is a long and distinguished history of nature writing in the English language that goes all the way back to Gilbert White's great 18th century classic work on the natural history of one small place. The three centuries since have seen a succession of observers of nature whose gifts in communicating the thrill of their experience is matched by their skilled use of the English language. Many of these have adopted the calendar as the template for their observations, among them L.J.F. Brimble and Edward Step, whose name – echoed in the title of Rachel's new book – reflects her acquaintance with all these forebears.

One thing Rachel shares with the great Gilbert White is that, if you turn the pages of her book at random, you never know what you are going to find: the entire spectrum of life on land – at least in our small corner of the world – seems to be here. Her geographical remit is not quite as narrow as White's Selbourne, the parish in which he was born and died in the same house, but it still focuses on a small number of key focal locations within reach of her home outside Birr.

Buzzard,
Buteo buteo

An extraordinary difference between Rachel and all her literary mentors, however, is that their nature calendars are the work of a lifetime, whereas she fills her pages with the crowded encounters of only a few years.

Rachel writes in a direct, simple, style that makes what she describes accessible to the widest range of readers, avoiding the technical language that so often smothers the excitement of discovery. And her words are balanced by pictures that are sometimes stunning: not only her photographs, but the accompanying artwork. A nature diary with such scope and with such a wide appeal is almost unique. In its pages, it conveys the excitement of first encounters with a cross-section of the amazing plants and animals that are to be found just over the threshold of our everyday lives. It will appeal as much to experienced naturalists – few of whom cast their taxonomic nets of discovery so widely – as it will open doors for a new generation of neophyte naturalists.

Step into Nature is an inspiring account of direct encounter with the everyday flora and fauna we can all see for ourselves, with our own eyes, once we learn to stop and look, an experience enriching beyond anything we can see on our screens, in the only corner of the world we can actually engage with and preserve for future generations.

John Feehan

Good Friday Grass, *Luzula campestris*

General note

This book is a personal awakening and appreciation of the astounding beauty and complexity of nature around us, in our gardens, hedgerows, woodland walks and bogs. It has evolved over a five-year period, from a general awareness to a thirst for more information, greatly helped by so many patient experts and a key selection of wonderful books. All photographs were taken with just a smartphone and as with all sketches and paintings, are my own. The book is set out as a weekly diary to provide a chronological view of some of the finds we can look out for in the weeks and months ahead. It covers, briefly, wildflowers, insects and other invertebrates, and fungi, generally found in the garden or on local walks.

The book is based on the superb publications produced in the late nineteenth century and early twentieth century, by nature writers such as Edward Step or Anne Pratt. Step, in particular, has a broad range, from detailed books based on trees or wildflowers to the incredibly useful and inspiring

Red-tailed Bumble Bee, *Bombus lapidarius* on Common Knapweed, *Centaurea nigra*

Step into Nature

Nature Rambles and *Nature in the Garden*. These books look at a diverse range of flora and fauna as they appear throughout the year.

This publication seeks to emulate that undertaking, highlighting the vast range of incredible species to be found locally. While set in Offaly, most highlighted species can be found throughout Ireland. Each plant or insect has been recorded on the National Biodiversity Data Centre (NBDC) allowing a collection point for individuals' records, while assisting with national monitoring. It is not a technical or scientific document but seeks to provide interesting facts about a broad range of species. With each new discovery, we can add to our knowledge and in the words of Edward Step, '*In finding what we sought, we are almost sure to find in addition something that is new to us*', Step, E., *Nature Rambles: Winter to Spring*. By simply slowing down, listening and observing, we can discover miniature wonders of nature. Wonders which fill us with awe when we learn of their complexities. Or those that are so heartachingly beautiful that the memory stays with you, long after the seasons change. Nature is compulsive, always available and dynamic - we can never have too much of God's creations.

Step, E., *Nature Rambles: Winter to Spring* (Warne & Co., 1933)

General note

Worldwide, we are witnessing catastrophic biodiversity losses, so it is all the more vital for each individual and community to help in a small but meaningful way, to slow down these losses. We must pay attention to what is around us, record it, tell others, and where at all possible, make efforts to increase biodiversity in our gardens, workplace, schools and communities. Each day can provide the opportunity to listen to the voice of nature, to reconnect, instilling us with a sense of fulfilment. Looking up into the heavens for passing birds, or pausing by a hedgerow, peering into the leaves, allows the naturalist look for minute wonders. Each discovery is a small triumph and such encounters have extraordinary potency.

Sometimes their scale can be difficult to grasp and as a small guide, the line under each page number is 10mm, while that under the book title (top right) is 30mm. Footnotes are not applied, but Further Reading is provided for each week at the end of the book. The Index is set out with individual common names where in general use, and also general groups such as Plants, Moths, Butterflies, Hoverflies etc.

Sphaerophoria scripta, (5-7mm, June to August), dark form, August - W1

Step into Nature

Blue September Sky

A descent of finches
sunlit
on rosehips
flickers of green
glint
underwing

How they pierce
the ripened
flesh
with dedicated
purpose
feast on a gash
of yellow
seed-spill

Laura Mc Kenna

Goldfinch, *Carduelis carduelis*,
feeding on rosehips

Contents

January 18-29

Week 1: House Sparrow, Candlesnuff Fungus, Glistening Inkcap, Yellow Brain, Scarlet Caterpillarclub

Week 2: Maidenhair Spleenwort, Rustyback, Wall-rue, Polypody, Mexican Fleabane, Common Chickweed, Shepherd's-purse, Red Dead-nettle

Week 3: Green Shieldbug, 7-spot Ladybird, Fox, Common Frog

Week 4: Lesser Celandine, Coltsfoot, Snowdrop, Winter Aconite, Long-tailed Tit

February 30-39

Week 1: Whooper Swan, Gorse Shieldbug, Hairy Shieldbug, Gorse Weevil

Week 2: Devil's Matchstick, Cup Lichen, Hazel

Week 3: Common Frog, caddisflies; *Limnephilus*

Week 4: Barren Strawberry, Wild Strawberry, Opposite-leaved Golden-saxifrage, Herb-Robert, Ivy-leaved Toadflax

March 40-73

Week 1: Starling, Eyed Ladybird, 14-spot Ladybird, 7-spot Ladybird, Ground-ivy, Common Field-speedwell, Dog's Mercury, Alexanders

Week 2: Juniper Shieldbug, Common Mouse-ear, Common Whitlowgrass, Blackthorn, Willow, Buff-tailed Bumblebee, White-tailed Bumblebee, Common Carder Bee, Red-tailed Bumblebee, Garden Bumblebee, Early Bumblebee

Week 3: Dame's Violet and moth *Plutella porrectella*, Common Quaker, March Moth, Hebrew Character, Early Grey

Week 4: Water Cricket, Lesser Diving Beetle, Large-grooved Diving Beetle, Smooth Newt, Water Measurer, Green Tiger Beetle, Puss Moth, Toothwort, Greater Stitchwort, Brimstone, Peacock, Small Tortoiseshell, Comma

April   74-115

Week 1: Wood Anemone, Bank Vole, Wood-sorrel, Primrose, Cowslip, False Oxlip, Early Purple-orchid, Common Twayblade, Pineappleweed, Field Madder, moths; Double-striped Pug, *Ypsolopha mucronella*, Shoulder Stripe, Oak Beauty, Yellowhammer, Siskin

Week 2: Ashy Mining Bee, Orange-legged Furrow Bee, Early Mining Bee, Chocolate Mining Bee, *Nomada marshamella, Nomada goodeniana, Entomophthora* fungus, Lords-and-Ladies and Owl Midge, Good Friday Grass

Week 3: Orange-tip butterfly, Garlic Mustard, Cuckooflower, Green-veined White, Holly Blue, Red Admiral, Bog Rosemary, Bilberry, Round-leaved Sundew, Oblong-leaved Sundew, Great Sundew, Pignut, Cow Parsley and leafmine *Phytomyza chaerophylli*, Bluebells

Week 4: Wild Garlic and Ramsons Hoverfly, hoverflies; *Rhingia campestris, Eristalis pertinax, Heldophilus pendulus, Epistrophe eligans, Syritta pipiens, Eupeodes luniger*, Large Red Damselfly, Pine Ladybird, Heather Ladybird, Orange Ladybird, Cream-spot Ladybird, Two-banded Longhorn Beetle, Marsh Marigold and moth *Micropterix calthella*

Contents

May   116-167

Week 1: Hedgehog, Red Squirrel, Rabbit, Emperor Moth, Pale Tussock, Dark Tussock, Garden Tiger, Pebble Hook-tip, Scalloped Hook-tip, Oak Eggar, Pebble Prominent, Grass Emerald, Bogbean, Bog Stitchwort, Common Butterwort, Field Forget-me-not, Bush Vetch, Common Vetch, Woodruff, Flower Crab Spider, Noon Fly, conopids; *Sicus ferrugineus*, solitary bees; *Andrena praecox, Andrena wilkella,* Blood Bee *Sphecodes* sp., Four Coloured (Forest) Cuckoo Bee

Week 2: Hawthorn, Small White, Wood White, Dingy Skipper, Small Copper, Green Hairstreak, Common Blue, Bird's-foot-trefoil, Wood Avens, Lousewort, Water Avens, cranefly; *Tanyptera atrata,* Birch Leaf Roller, sawflies; *Euura pavida, Cimbex femoratus,* Four-spotted Chaser, Hairy Hawker, Common Blue Damselfly, Variable Damselfly, Azure Damselfly, Blue-tailed Damselfly, Narrow-bordered Bee Hawkmoth, Common Purple and Gold, Small Purple-barred, Burney Companion, Swift, Swallow, Sand Martin, House Martin

Week 3: Sanicle, Mountain Everlasting, Tormentil, Creeping Cinquefoil and leafmine *Fenella nigrita,* Yellow Corydalis, Yellow-rattle, galls; *Dasineura ulmaria, Acalitus longisetosus, Jaapiella veronicae,* Common Cockchafer, Common Pill Beetle, Red-headed Cardinal Beetle, Malachite Beetle, mirid bugs; *Liocoris tripustulatus, Heterotoma planicornis, Stenodema laevigata, Harpocera thoracica,* Birch Catkin Bug, Spotted Lacehopper, hoverfly *Leucozona lucorum,* sawfly *Aglaostigma aucupariae,* mayfly *Cloeon dipterum*

Week 4: Eyelash Fungus, Fine-leaved Sandwort, Quaking-grass, Burnet Rose, Bloody Crane's-bill, Parent Shieldbug, Bronze Shieldbug, Spiked Shieldbug, Ruby-tailed Wasp, *Ancistrocerus gazella,* beetles; *Malthinus flaveolus, Malthodes marginatus,* Scale Insect, Fox Moth, Bagworm, *Pschye casta,* Marsh Fritillary

June 168-217

Week 1: Blue Shieldbug, beetles; *Cassida rubiginosa, Cassida vibex,* mirid bug; *Grypocoris stysi,* Lesser Thorn-tipped Longhorn Beetle, Marmalade Hoverfly, Potter Wasp, *Ancistrocerus trifasciatus,* sawflies; *Pristiphora conjugata, Pristiphora leucopus, Abia nitens, Eriocampa ovata, Nematinus steini,* Common Figwort, Figwort Weevil, Goat's-beard, Bee Orchid, Poplar Hawk-moth, Ghost Moth, Chinese Character, Angle Shades, Diamond-back, Nettle-tap, Marbled White Spot, Six-spot Burnet

Week 2: Small Heath, galls; *Taphrina pruni, Eriophyes similis,* hoverflies; *Chrysotoxum bicinctum, Scaeva pyrastri,* Green-winged Orchid, Early Marsh-orchid, Pyramidal Orchid, Lesser Butterfly-orchid, sawflies; *Cimbex conatus, Strongylogaster multifasciata, Heterarthrus nemoratus, Hemichroa crocea, Allantus* sp., caddisflies; *Limnephilus lunatus, Limnephilus auricula, Glyphotaelius pellucidus, Mystacides azurea, Oecetis ochracea, Phryganea* sp.*, Stenophylax permistus, Hagenella clathrata,* alderfly; *Sialis* sp.

Week 3: Marsh Cinquefoil, Royal, Fern, Meadow Thistle, Bog Asphodel, Lesser Water-plantain, Common Valerian, Skullcap, Bog Pimpernel, Ragged Robin, Clouded Drab, Magpie, Ringlet, Large White, Meadow Brown, Painted Lady, Mottled Grasshopper, mirid bugs; *Campyloneura virgula, Dicyphus errans, Leptopterna dolabrata, Capsus ater, Heterocordylus tibialis, Ditropis pteridis,* Pine Marten, Irish Stoat, Otter

Week 4: Swallow-tailed Moth, Light Emerald, Peppered Moth, Burnished Brass, Lilac Beauty, Spectacle, Grey Arches, Dot Moth, Muslin Footman, Elephant Hawk-moth, Eyed Hawk-moth, Ruby Tiger, hoverflies; *Xylota segnis, Volucella bombylans, Eristalis horticola,* 14-spot Ladybird and wasp *Dinocampus coccinellae,* Banded Demoiselle, Beautiful Demoiselle, Puss Moth, Four-banded Longhorn Beetle, Garden Chafer

Contents

July 218-253

Week 1: Brown Hawker, Emerald Damselfly, Ruddy Darter, Common Darter, Black-tailed Skimmer, Common Froghopper, Field Scabious, Cow-wheat, Common Centaury, Eyebright, Garden Bumblebee, wasp *Ectemnius lapidarius*, hoverflies; *Leucozona laternaria, Leucozona glaucia, Volucella pellucens*

Week 2: Great Willowherb and mirid *Dichypus epilobii*, Purple-loosestrife, moths; *Argyresthia brockeela, Argyresthia goedartella*, Hedge Woundwort, Marsh Helleborine, Red Bartsia, Beautiful China-mark, Brown China-mark, Forest or Red-legged Shieldbug, Twin-lobed Deerfly

Week 3: Moths Bordered Beauty, Purple Clay, *Nemophora minimella*, Small Chocolate-tip, Hemp-agrimony Plume and Hemp-agrimony, Yellow-wort, Water Mint, Wild Marjoram, Treecreeper, Yellow-bodied Black Fungus Gnat, beetles; *Ampedus* sp., *Ctenicera cuprea*

Week 4: Fungi *Ergot*, mirid bug *Calocoris roseomaculatus*, Parent Shieldbug, Shark (moth), Mother of Pearl, Canary-shouldered Thorn, *Psychoides filicivora* and Hart's-tongue, Blue Fleabane, hoverfly *Eristalis intricaria*, Ichneumon wasp, Common Knapweed and moth *Cochylimorpha straminea*, Knapweed Gall-fly and *Urophora quadrifasciata*, Chalcid wasp

August 254-291

Week 1: Hoverflies; *Sphaerophoria scripta, Sericomyia silentis*, gall *Kiefferia pericarpiicola*, Wild Carrot, mirid bug *Apolygus spinolae*, Tortoise Bug, sawfly *Caliroa cerasi*, Wild Angelica, Gypsywort, Knotted Pearlwort, Ear Moth, Haworth's Minor, Grey Scalloped Bar, Gold Spot

Week 2: Silver-washed Fritillary, Sharp-tail Bee, leafhoppers; *Evacanthus interruptus, Cicadella viridis, Zonocyba bifasciata*, galls;

Robin's Pincushion, Alder Tongue, *Iteomyia major, Pemphigus bursarius*, Scarce Blue-tailed Damselfly, Trifid Bur-marigold, Autumn Gentian, Field Gentian, Soapwort, Leafcutter Bee

Week 3: Common Green Grasshopper, beetle *Donacia* sp., wasp *Hepiopelmus variegatorius, Cotesia glomeratus* and Large White Butterfly caterpillar, Hieroglyphic Ladybird, 2-spot Ladybird, mirid bugs; *Drymus sylvaticus, Pithanus maerkelii, Malacocoris chlorizans*, White Plume Moth, sawflies; *Euura ferruginea, Arge ustulate*

Week 4: Stiltbug *Metatropis rufescens* and Enchanter's-nightshade, Sputnik Spider, Candy-striped Spider, hoverfly *Eupeodes corollae*, mirid bugs; *Dicyphus stachydis, Dicyphus pallicornis, Pinalitus cervinus, Deraeocoris lutescens*, Common Hawker, Migrant Hawker, Grass-of-Parnassus, Autumn Lady's-tresses, sawflies; *Tenthredo colon, Euura papillosa, Cladius grandis, Diprion pini*, Plasterer Bee *Colletes succinctus*, Red Clover and micro-moth *Coleophora deauratella*

September 292-329

Week 1: Sallow, Large-flowered Evening-primrose, Traveller's-joy, Bladderwort, Common Fleabane, mirid bug *Pantilius tunicatus*, hoverfly *Platycheirus granditarsus*, Field Digger Wasp *Mellinus arvensis*, Small Scabious Mining Bee, Spiders Zebra Jumping Spider, *Heliophanus* sp., Garden Spider, *Arctosa perita*, Four-spotted Orbweb Spider, Large Marsh Grasshopper, Black Darter, Slender Groundhopper, Common Groundhopper

Week 2: Mirid bugs; *Lygus wagneri, Adelphocoris lineolatus*, Potato Capsid Bug, *Plagiognathus chrysanthemi, Plagiognathus arbustorum*, Perforate St John's-wort, Fairy Flax, galls; *Hartigiola annulipes*, Fig Gall, *Chirosia grossicauda*, sawflies; *Platycampus luridiventris, Eutomostethus ephippium, Fenusella nana, Cladius brullei*

Contents

Week 3: Clouded Yellow, The Vestal, Rusty-dot Pearl, Vapourer, Black Rustic, mirid bug *Adelphocoris seticornis*, hoverfly *Eristalis intricaria*, Field Grasshopper, Pea Gall, Smooth Spangle Gall, Silk Button Gall, Common Spangle Gall, Common Puffball, Stinkhorn, Fairy Inkcap, Sulphur Tuft, Green Elfcup, Shaggy Parasol

Week 4: Barn Owl, Grey Wagtail, Dipper, leafmines; *Phyllonorycter coryli, Parornix devoniella, Stigmella hemargyrella, Stigmella tityrella, Stigmella plagicolella, Phyllonorycter spinicolella, Stigmella oxyacanthella, Phyllonorycter oxyacanthae,* Fly Agaric, Yellow Stagshorn, Shaggy Ink Cap, Deceiver, wasp *Dyscritulus* sp., caddisfly *Halesus radiatus,* beetle *Sphaeridium* sp.

October   330-353

Week 1: Humming-bird Hawk-moth, Green-brindled Crescent, Hawthorn, Crab Apple, Guelder Rose, Blackberry, Blackthorn, Bush Vetch, Wild Rose, Ivy, Spindle, Collared Earthstar, White Saddle, Elfin Saddle, Daisy

Week 2: Figure of Eight, Lunar Underwing, Grey Shoulder-knot, Red-line Quaker, Blair's Shoulder-knot, Papillose Bog Moss, Red Bog Moss, Feathery Bog Moss, Lustrous Bog Moss, Broom Fork Moss, Pointed Spear-moss, Ichneumon wasps; *Campodorus* sp., *Mesochorus* sp., shieldbugs; Birch, Bronze, Green, Hairy and Hawthorn

Week 3: Penny Bun, Oysterling, leafmines; *Leucoptera laburnella, Leucoptera lotella, Acrolepia autumnitella, Phyllonorycter nigrescentella,* Broad Damselbug, Marsh Damselbug, Thread-legged Bug, Speckled Wood

Week 4: Dark Honey Fungus, Field Bird's Nest, Pestle Puffball, Blue Roundhead, Clouded Funnel, Sulphur Tuft, Tawny Funnel, Deceiver, Parrot Waxcap, slime moulds; *Trichia varia, Ceratiomyxa fruticulosa*

Step into Nature

November — 354-375

Week 1: Fallow Deer, galls; *Andricus kollari, Andricus curvator,* Knopper gall, Snipe, Purple Toadflax, Lemon Disco, Purple Jellydisc, Coral Spot, Upright Coral,

Week2: *Ichneumon sarcitorius,* The Sprawler, Feathered Thorn, Grey Heron, Parrot Waxcap, Lilac Fibrecap, Wood Blewit, Amethyst Deceiver

Week 3: Leafmines; *Ectoedemia subbimaculella, Stigmella luteella, Stigmella betulicola, Ectoedemia argyropeza,* Buzzard, slime mould *Metatrichia floriformis,* Crystal Brain, Beech Jellydisc, Wrinkled Club

Week 4: Stoneflies; *Leuctra fusca, Isoperla grammatica, Protonemura meyeri,* moths; Brick, *Mompha subbistrigella,* Lapwing, Golden Plover, Mallard, Moorhen, Stonechat

December — 376-391

Week 1: Jellybaby, Holly Leaf Gall Fly, Holly Speckle, Holly Parachute, Great Tit, Blue Tit, Coal Tit, Greenfinch, Dunnock, Goldfinch, Wren

Week 2: Ants; *Formica* sp., *Myrmica* sp., *Lasius* sp., 22-spot Ladybird, 10-spot Ladybird, Common Earwig, Lesne's Earwig, Lesser Earwig, Pill Millipede

Week 3: Mirid bug *Chilacis typhae,* spiders; *Neriene montana, Pachygnatha clercki, Zygiella x-notata,* Great Pond Snail, Brown Lipped Snail, Strawberry Snail, Glass Snail, Amber Snail, Garden Snail, Sieve-winged Snail Killer

Week 4: Raven, Rook, Jackdaw, Hooded Crow, Winter Moth, December Moth

Acknowledgements and Further Reading — 392-409

Index — 410-431

January

Week 1 January is generally a cold, grey month in Ireland with wind, rain and uncertain temperatures. But it is also an opportunity to see the bones of the earth, to take notice of tree or shrub locations and start planning for spring and summer in the garden. With warm, waterproof clothing, it is worth snatching any available time, during daylight hours to take a short walk. We can make up for the scarcity of flowers or insects by

looking for fungi still remaining, or the many small birds visiting the garden or hedgerows in search of food, like the reliable House Sparrow.

There are numerous fascinating fungi nestled in fallen leaves. Candlesnuff Fungus may be found all year-round on dead wood along with the aptly named Glistening Inkcap, Yellow Brain, *Tremella* sp. is parasitic on *Peniophora*, which is a crust fungus found on Gorse and Hazel.

House Sparrow, *Passer domesticus*, (14-16cm, all year)

Step into Nature

Candlesnuff Fungus, *Xylaria hypoxylon*, (up to 6cm high, all year)

Glistening Inkcap, *Coprinus micaceus*, (stipe 10cm, cap 3cm across, March to January)

Yellow Brain, or Witch's Butter, *Tremella sp.* (fruiting body up to 6cm, all year)

Scarlet Caterpillarclub, *Cordyceps militaris*, (6cm high, August to January), garden

January - Week 2

Scarlet Caterpillarclub, *Cordyceps militaris*, is a parasitic fungus which grows out of buried moth pupae. We were fortunate enough to discover one in the garden, close to the short grass, with its vibrant colour and speckled appearance. *Cordyceps* fungi are a fascinating group - for further information relating to their different species, refer to *Entangled Life*, by Merlin Sheldrake.

Week 2

An old stone wall is always worth examining in a town or the countryside. Stone walls provide islands mapped with lichen and a good range of common ferns, visible all year. Maidenhair Spleenwort is a delicate fern with shiny black stalks and small green leaflets. Rustyback is so named because the underside of the fronds are covered with soft silvery scales. As the fern ages, these become rust-coloured. Wall-rue has divided fronds and was once used as a remedy for rickets. Polypody, an attractive fern, can cover large areas of wall or trees, tufting out from the mossy cracks. Spores found on the underside of certain ferns will be examined later in the year for the presence of micro-moths (see July - W4).

Step into Nature

Maidenhair Spleenwort, *Asplenium trichomanes*,
(4-20cm long, all year, spores May to October)

Rustyback, *Asplenium ceterach*,
(3-12cm long, spores April to October)

Wall-rue, *Asplenium ruta-muraria*,
(10cm long, spores June to October)

Polypody, *Polypodium* sp.,
(30cm long, spores August to March)

January – Week 2

Common Chickweed, *Stellaria media*, (30cm high, flower 5-10mm)

Shepherd's-purse, *Capsella bursa-pastoris*, (35cm high, flower 2-3mm)

Red Dead-nettle, *Lamium purpureum*, (30cm high, flower 18mm, all found throughout the year, garden

Step into Nature

A few hardy plants may be found at the start of January. It very much depends on the general temperature at the time, with some flowering a month earlier than during a colder year. Mexican Fleabane, *Erigeron karvinskianus*, is a resilient, fast-spreading plant. It can grow in the smallest of pavement cracks or fill an entire flower bed, once it puts down roots under the rays of the sun.

The small flowers of Common Chickweed with their five deeply notched petals and red stamens add a splash of much needed colour across gardens and fields. Common Chickweed is said to have been boiled with red rose leaves and oil to relieve cramps and the seeds are popular among small birds. Another abundant wild plant growing in fields, gardens or waste ground is Shepherd's-purse. Its name comes from the heart-shaped seed-vessels, thought to resemble old leather purses. Another attractive flower at this cold time of year is Red Dead-nettle, with its pale-purple, labiate-shaped flowers. It continues to flower throughout the year.

Mexican Fleabane, *Erigeron karvinskianus*, (50cm high, 16mm flower, all year, garden)

January - Week 3

Week 3   As occasional blue skies and increased bird song signal approaching spring, this season provides an opportunity to admire the many colourful mosses in their full beauty during these cool, damp months (see also October - W2).

Ladybirds and shieldbugs, such as the Green Shieldbug, which have overwintered, may be found within tree branches. The joints of railings or headstones in graveyards are a popular location for ladybirds in particular. This 7-spot Ladybird was found on Scots Pine in Killaun Bog, where the shy fox was also glimpsed, peeping through the long grass. Take care if clearing fallen leaves in the garden, as they are often home to hibernating frogs who will make their way to their preferred pond or stream to mate, towards the end of February (see February - W3).

Green Shieldbug, *Palomena prasine*, (12-13.5mm) and 7-spot Ladybird, *Coccinella septempunctata*, (6-8mm, all year garden)

Common Frog, *Rana temporaria*, (6-9cm long, garden)

Step into Nature

Fox, *Vulpes vulpes*, peeping through long grass at Killaun

Adults (c1.5m nose to tail), mate December to February, cubs (up to 14) March to April

January - Week 4

Lesser Celandine, *Ficaria verna*, (up to 20cm high, flowers 20-30mm, January to April)

Coltsfoot, *Tussilago farfara*, (10-30cm high, flowers 15-30mm, January to April, garden)

Snowdrop, *Galanthus nivalis*, (7-15cm high, January to March), Birr Demesne

Winter Aconite, *Eranthis hyemalis*, (15cm high, 2-3cm flowers, January to February), Birr Demesne

Step into Nature

Week 4

The great charm of the impending spring is watching for much-loved flowers. Few sights can be more cheerful in the coming months than an expanse of shining Lesser Celandine a true celebration of spring. The bright, yellow flowers, related to the Buttercup, open from January to May and found in damp grassy places. As temperatures begin to rise, Coltsfoot sends up their hollow, cottony stems clad with large scales and topped with a vibrant flower head. Similar to the Daisy, the flower head is actually made up of closely packed, tiny individual flowers or florets. These later develop into a 'clock' type seed head, like the Dandelion. It is only when the flowers are over that their heart-shaped leaves appear. The leaves were used as an infusion for coughs and the scientific name is derived from *tussis*, a cough; while the upper surface, covered with a thick cottony down, was used as tinder.

The delicate Snowdrop is always a pleasure to behold, sometimes gracing snow covered banks. While referred to as a garden flower, it can spread extensively through woods intermingling with Lesser Celandine producing the most picturesque effect. Winter Aconite with its cheery yellow flowers encircled with an elegant green collar, may also spread from gardens to enliven the month of January.

January - Week 4

A chance encounter can fill your day with joy and what could be more delightful than an animated flock of Long-tailed Tits? The small black and white birds have a fluffy belly suffused with pink, and can be found around areas of deciduous woodland or mature hedgerows. They are pure perfection.

Their extremely long tail is particularly noticeable in flight. We were thrilled to spot the busy group of seven pecking the ground for small insects in a garden in Tullamore.

Classed as a common resident species, they can be found in all counties in Ireland. During the winter, they gather in large flocks with other tit species.

Long-tailed Tit, *Aegithalus caudatus*, (13-15cm long including 7mm tail), all year, garden

Step into Nature

They construct the most beautiful and ingenious nests, built in thorny bushes such as Hawthorn or Gorse for protection against predators. Bottle or ball-shaped, the nest is made up of feathers and spider webs, while the outside is camouflaged with moss and lichen. The spiders' silk allows the nest to expand as the hatched chicks grow.

Some nests are said to contain over 2,000 feathers in their construction.
The Long-tailed Tits form sociable groups with aunts and uncles helping to feed and raise chicks.

As a group, they are very vocal and will keep up a continuous chatter in search of food or materials for nest building. They eat insects, seeds and nuts and will take peanuts and fat at bird tables. Such bird tables are particularly important during the winter months to help small birds struggling to find food.

Nest of Long-tailed Tit,
Aegithalus caudatus

February

Week 1 February announces the coming of spring and, while still cold, it does allow for some walks further afield. A visit to Turraun Bog with blue skies chased by black clouds provides an opportunity to see some hardy insects enjoying the early sunshine. A large flock of Whooper Swans, with their yellow bills, honk loudly on the lake, as they visit us from Iceland.

Looking at Gorse at this time of year reveals two different shieldbugs which are sheltering on the plant. They are members of the Hemiptera order which includes aphids and planthoppers. The Gorse Shieldbug lays its eggs in two diagonal rows, usually along a Gorse bud. The hatched nymph goes through five instar changes as it develops into adulthood, and the large adult has two colour forms. Predominantly green as they emerge in the spring, the late-summer generation has purple-red markings which darken prior to overwintering.

Whooper Swan, *Cygnus cygnus*, (140-165cm long, 200-275cm wingspan, October to April)

Step into Nature

Gorse Shieldbug eggs in April, Killaun Bog

Gorse Shieldbug nymph (April to August) in July, Cranberry Bog, Gallen

Gorse Shieldbug, *Piezodorus lituratus*, a late-summer adult in February

Gorse Shieldbug, *Piezodorus lituratus*, (10-132mm, all year). a spring adult in August

February – Week 1

A Hairy Shieldbug, also found on Gorse, does not shy away from the camera, like the more bashful Gorse Shieldbug. As its name would suggest, it is covered with long hairs, particularly noticeable on the furry nymphs. The antennae and connexivum (abdomen border) have a striking banding of black and white.

On the same Gorse plant, a number of tiny Gorse Weevils were found, digging into soft tissue of stem and spines with their snouts, creating characteristic round holes as evidence. Their larva emerge from the egg inside the Gorse seed pod and feed on the seeds for six to eight weeks. Weevils (known for their elongated snouts) are beetles, with approximately 214 species in Ireland.

Hairy Shieldbug adult, *Dolycoris baccarum*, (11-12mm, all year)

Hairy Shieldbug nymph, *Dolycoris baccarum*, (June to September), August, Cranberry Bog, Gallen

Gorse Weevil, *Exapion ulicis*, (2-3mm, all year), Turraun Bog

Step into Nature

Week 2 Many varieties of the enduring organisms known as lichens are visible during the winter, instilling a sense of timelessness.

Devil's Matchstick, *Cladonia* sp., (up to 3cm high), Killaun Bog

They form as the result of a symbiotic association between fungi and algae. Lichen occur in three main forms; crustose (thin crust), foliose (radiating patches) and fruticose (miniature bushes). Devil's Matchstick are found on peaty soil, with grey stalk, scaly base and vibrant red spore-producing tips. Cup lichens (a combination of foliose and fruticose) form mats of grey-green scales which send up stalks shaped like long-stemmed wine glasses. These can be found filled with dew drops and interlaced with jewelled webs on decaying timber and stone walls.

Cup lichen, *Cladonia* sp., Lough Boora

February - Week 2

It is such a promising sight to glimpse the long, swaying yellow tassels (catkins) of Hazel, brightening a woodland walk and providing a clear indication of the location of these wonderful trees. The catkins first appear towards the end of the year, shorter and green or grey in colour. Through the winter, they continue to slowly lengthen, turning bright yellow, their scales separate revealing the tiny male flowers, approximately 100 per catkin. For the Hazel is monoecious, with separate male and female flowers on the same tree.

As Hazel flowers during an Irish February, it devised a plan to keep the flowers and their pollen dry. By hanging its catkins downwards, the bract (modified leaf) acts as an umbrella, keeping the pollen away from rain. The same branch may also hold, at a deliberately higher level, the miniature female flower, which would pass as a bud. The vibrant red threads are the styles (stalk) of the pistils (reproductive part), hoping to catch pollen from another Hazel.

Hazel, *Corylus avellana*, male catkins, (5-12cm)

Step into Nature

Hazel, *Corylus avellana*, female flower, (1-3mm)

February – Week 3

Week 3

On a mild day by the garden pond, streams or bog pools, the guttural gurgle of the Common Frog can be heard. Males arrive first and with a pair of vocal sacs produce their love song to attract females, resulting in what was discretely described by my Mother as 'lots of activity in the pond'. The female lays 1-2,000 black eggs at the bottom of the water, while the male releases sperm. The eggs are fertilised immediately, and as their surrounding jelly absorbs water, they swell to four times their original size, causing them to rise to the surface.

Both male and female frogs return to the same pond year after year. Interestingly, they do not feed on aquatic creatures but on slugs, insects, worms, and spiders. We will keep an eye out for the tadpoles as they hatch and grow from April to May, leaving the pond in June or July as tiny froglets.

Step into Nature

Caddisfly case of reeds (Killaun), bi-valve shells (Turraun) and twigs (garden)

It is a good time of year to look for evidence of Caddisflies (Trichoptera). They begin life within bog pools, ponds or rivers and as soft larvae. For their protection, many create intricate cases out of surrounding plants, twigs, stones or shells. Size is no measure of complexity or beauty.

These ingenious homes allow the tender larva move about the water in relative safety. They first weave a silken tube, gradually attaching available material. The more sturdy head and legs protrude from the front, allowing for movement through the water. *Limnephilus* sp. avail of delicate waterweeds as shown here. As the larva develops and expands in size, it cuts a piece off the rear of the case and makes additions to the front. We will look for the moth-like adults later in the year, June - W2.

February – Week 4

Week 4

Towards the end of February, more flowers may be found, providing a dash of colour and interest. The Barren Strawberry, as its name suggests, does not produce the bright red fruit of the Wild Strawberry. But it does provide a beautiful delicate flower with five white, lightly notched petals widely separated at the base.

The flower of the Wild Strawberry has five rounded petals and the rich, diminutive fruit that adds refreshment to a woodland walk. It flowers from April, with fruit from June. The botanical name is due to the sweet odour while the common name may come from the old practice of threading the fruits on a straw as cultivated for sale in England since the 15th century.

Barren Strawberry, *Potentilla sterilis*, (7-15cm high, flower 15mm, February to June)

Wild Strawberry (July), *Fragaria vesca*

Wild Strawberry (May), *Fragaria vesca*, (5-10cm high, flower 20mm, April to July)

Step into Nature

Opposite-leaved Golden-saxifrage, *Chrysosplenium oppositifolium*, (15cm, flower 3-5mm, February to July)

Herb-Robert, *Geranium robertianum*, (40cm high, flower 12-18mm, February to October)

Ivy-leaved Toadflax, *Cymbalaria muralis*, (5cm high, flower 8-15mm, Feb to Nov)

On damp ground with a tang of moss, you may notice a carpet of tiny green and yellow flowers, giving the impression of sun-dappled light. It is Opposite-leaved Golden-saxifrage, with a disproportionately large name.

Could there be anything more beautiful than a fleeting, panoramic view of the world from a single raindrop? The hardy and prolific Herb-Robert produces flowers until November. Of the Cranesbill family, it has pink satin petals with three pale streaks. The name comes from the developing fruit after the petals fall, the 'bill' springs open to scatter the seeds, fascinating to view under a microscope.

Ivy-leaved Toadflax, found in the smallest crevices, tumbling over old walls, has small lilac (or white) flowers.

March

Week 1   A bright fresh morning in early March is filled with the chattering sound of numerous Starlings. A common resident, they are joined late-Autumn with influxes from the Continent. They nest in holes in buildings or trees and feed in large noisy flocks on insects, grain and berries. An audacious bird that walks with a swagger and has a surprising capacity to mimic both animate and inanimate sounds. The dark summer plumage has an exquisite array of blue, green and violet, iridescent in the bright sunshine.

This month brings an extraordinary increase in plant and insect activity, in particular over the last two weeks with moths, bees and frogs emerging for the spring.

Starling, *Sturnus vulgaris*, (20-23cm long, all year)

There are many ladybirds to be found in the garden, on bogs or in woodlands, generally emerging from hibernation in the spring. Females lay their eggs after mating, which hatch within a few days. The larvae moult three times over 2-4 weeks, pupate and emerge as adults a few weeks later.

Eyed Ladybird, *Anatis ocellata*, (7-8.5mm); 14-spot Ladybird, *Propylea quattuordecimpunctata*, (3.5-4.5mm)

The large and stunning Eyed Ladybird was found in the garden, despite normally associating with conifer woodlands. Up to 8.5mm, it is larger than the more common 7-spot and twice the size of the yellow and black 14-spot. The 18 black spots on the Eyed Ladybird are ringed with yellow, making it easily distinguishable. Its large pupa is seen resting on a Birch leaf. A welcome addition to any garden as both adults and larvae feed on aphids.

Eyed Ladybird, *Anatis ocellata* and 7-spot Ladybird, *Coccinella septempunctata*, (5.5-8mm)

Eyed Ladybird pupa, *Anatis ocellata*, (August), all found in the garden

March – Week 1

We can expect to find the bright blue-purple flowers of Ground-ivy as soon as Coltsfoot has flowered. While not related to Ivy, it creeps along the base of hedgerows, in a similar manner. It has kidney-shaped leaves and a combination of trailing stems and erect branches which bear the tubular flowers with a broad lower lip (where insects land). For such a small plant, it has had an incredible amount of purported uses, from assisting with hearing aliments to inflammation of the eyes, both human and animal.

The tiny bright-blue flowers of Common Field-speedwell cover areas of bare or waste ground. Similar to other members of the Speedwell family, this four-petal flower is distinguished by its paler lower lobe.

Ground-ivy, *Glechoma hederacea*,
(5-25cm high, flower 15-25mm,
March to September)

Common Field-speedwell,
Veronica persica,
(c.20cm high, flower 8-12mm, all year)

The green leaves and flowers of Dog's Mercury and its low-growing habit, leave the plant relatively inconspicuous in the few woods where it thrives. It does not need showy flowers to attract insects as it is wind pollinated and dioecious, with separate male and female plants. These are found in Birr, as part of Lord Rosse's woods.

Another predominantly green plant is the taller, vigorous Alexanders with its shiny leaves and stout hollow stems. The yellow umbels (flower cluster) are very popular with a selection of insects on a warm sunny day and it continues to flower into June. Stems were formerly used as a type of vegetable, in lieu of celery, while seeds were smoked with tobacco.

Dog's Mercury, *Mercurialis perennis*, (15-40cm high, March to May)

Alexanders, *Smyrnium olusatrum*, (1m high, flower 4-6mm, February to June), with Green-veined White butterfly, May

March – Week 2

Week 2

If ever you were to pick a manageable group of insects in a bid to seek out each of its species, it would have to be Shieldbugs (Hemiptera), or perhaps Ladybirds (Coleoptera), or maybe Dragonflies (Odonata). Whichever you choose, Shieldbugs provide fascinating variety with many conveniently located in gardens and parks. This stunning Juniper Shieldbug, the first recorded in Offaly, was found on Lawson's Cypress in Crinkill cemetery, 2022. Added to the Irish list in 1995, they have slowly spread across parts of Ireland. They are very distinctive with red boomerang-shaped markings, designed to blend in with their habitat.

Juniper Shieldbug, *Cyphostethus tristriatus*, (9.5-10mm, adult September to June, nymph June to September)

Step into Nature

Common Mouse-ear, *Cerastium fontanum*, (10-30cm high, flower 5-7mm, March to November)

Common Whitlowgrass, *Erophila verna*, (20c,10cm high, flower 3-6mm, February to May)

There is something very endearing about the perennial Common Mouse-ear, which, like annual Common Chickweed, grows freely across waste ground or fields. So much thought and care has gone into forming their perfect little flowers with five deeply-cleft white petals and central spiraling sepals. The hairy stems and pointed leaves in opposite pairs are thought to resemble the ear of a mouse.

Common Whitlowgrass is so small that it can escape notice as tiny clusters form in damp spaces on stone walls or paving. It is worth seeking out to appreciate the cross-shaped, pure white flowers rising out of slender downy stems from a rosette of pointed leaves at its roots. It was formerly mixed with milk as a cure for whitlow (infection of the finger or toenail).

March – Week 2

March is a month of many seasons in Ireland, with vast temperatures swings. We have to be ready to dash out at a moment's notice, when the sun comes out, bring a raincoat. When it is raining, start driving, as it will most likely have stopped, momentarily, when you arrive at your preferred location.

It is the month which brings an explosion of white in our hedgerows and across field boundaries with a profusion of impossibly delicate and incredibly beautiful flowers against a backdrop of dark, spinney, leafless branches. Blackthorn bows against prevailing winds but doggedly continues to grow. Their flowers, and those of Willows at this early stage in the year, are phenomenally important for emerging insects such as moths, butterflies, hoverflies and especially for queen bees who are desperately seeking sugar-rich nectar after a long, eight months hibernation. The thorny low-growing shrubs provide an excellent stock-proof barrier, although its effectiveness is considerably diminished by annual flaying with hedge-cutting machines. These cut back the young thorny, flowering shoots, leaving at the base the older stems which have developed without thorns. A more moderate three-year programme, cutting alternating sides of each field, would allow for more natural growth and increased, valuable flowering.

Step into Nature

Blackthorn, *Prunus spinosa*, (flowers 10-15mm, March to May)

Peacock butterfly, *Inachis io*, (63-75mm wingspan, March to September)

March - Week 2

Willows, *Salix* sp., male (pollen)   Willows, *Salix* sp., female, (fruit and nectar)

We will look at Blackthorn again as it comes into leaf, to check for signs of galls (June - W2), sawflies (August - W1) or micro-moths in the form of leafminers (September - W4)). The other tree of such importance at this time of year is the Willow. Pussy Willow covers a range of the smaller *Salix* species which flower in early spring. Unlike Hazel, there are separate male and female trees (dioecious) and, like Blackthorn, they flower before leaves begin to bud. Males produce pollen and females fruit and nectar.

The male catkins from Goat Willow, *Salix caprea*, were thought to resemble cats due to their furry appearance prior to fully coming into flower. As they flower, they produce striking yellow anthers. The female flowers are silver-grey, almost spikey in appearance. They are very easy to grow and any spare patch of ground would greatly benefit from the beauty of these harbingers of spring. Bees and other insects swarm around the trees in great numbers until sunset, when their place is taken by nocturnal moths.

Some of the earliest bees to appear are the familiar Buff-tailed and White-tailed Bumblebees. It is the large 20mm starving queens who first emerge from most habitats, including gardens. They go to Willows for nectar (female flowers) then protein-rich pollen (male flowers) to develop their stored eggs.

Buff-tailed Bumblebee, *Bombus terrestris*, (queen 18-20mm, worker 13mm, March to October), on Aubretia, garden

White-tailed Bumblebee, *Bombus lucorum*, (queen 16-20mm, worker 12mm, March to October), garden

March - Week 2

Initially called Humblebees by Darwin due to their humming sound, the name remained into the 20th century, still referred to as such in the Step 1930s books. Bumble came from their impossibly undersized wings and their sometimes bumbling flight as they awkwardly attempt to take off.

White-tailed Bumblebee, *Bombus lucorum* agg., wobbly lift-off, Dandelion, April

As the queens get sufficient nectar and pollen, they begin to search for a suitable nesting site, usually commandeering an existing hole in the ground. The Buff-tailed Bumblebee sways low to the ground around clear patches of soil. It will explore former rodent chambers and create an internal, insulated nest out of available moss, grass or feathers. The finished ball-shaped nest has a central, tennis-ball sized cavity and a small hole for the queen to enter through. Some species nest under patios, in roofs or in bird nests. Within the central cavity, the queen makes a wax cup for herself, which she fills with honey (concentrated nectar). She then makes a pea-sized ball out of collected pollen and coats it in wax. Into this ball, she houses her first batch of 16 eggs, fertilised as they leave her body.

Sitting over the ball, she 'shivers' to warm and incubate the eggs, in a similar manner to a bird. She feeds herself from the wax cup during this process. Over the four days required for eggs to hatch, the queen may use her own weight in sugar, each day. The honey required is the equivalent to having visited a staggering total of 24,000 flowers.

The hatched grubs feed off their pollen ball as the queen leaves the nest to gather more. After two weeks, the grubs spin individual silk cocoons. Two weeks later, the female worker bees emerge. Meanwhile, the hard-working queen has laid a second batch of eggs - she no longer needs to leave the nest as these are cared for by the worker bees.

The workers head out of the nest and make short reconnaissance trips, familiarising themselves with the terrain, and more importantly, available food supply. The queen continues to lay batches of eggs while the workers, several hundred by July, make the required honey and pollen containers. The queen now lays both female (future queens) and male eggs.

Common Carder Bee, *Bombus pascuorum*, (queen 13-17mm, worker 10mm, March to October), on garden Cranesbill

March - Week 2

The only role of the small, charmingly scruffy males is to feed and mate with future queens, who will store the sperm internally. These new queens soon hibernate (June to October), just below ground-level, and the few resilient survivors start the entire process afresh the following spring. The original one-year old queen and her workers have completed their separate roles and die off by September. For detailed information relating to this process, refer to the fascinating *A Sting in the Tail* by Dave Goulson.

We will take this opportunity to look ahead to some of the bumblebees we can expect to meet over the coming months. The widespread Common Carder Bee nests in grass or bird nests. It may be variable and confused with Large Carder Bee (no dark hairs on abdomen).

Beautiful, velvet Red-tailed Bumblebees nest under stones or in meadows. They are found in flower rich habitats and are subsequently becoming rarer. The medium sized Garden Bumblebee (July - W1) nests underground, have a long 'horse-like' face and are found in most habitats. Early Bumblebees, the incredibly cute small bees, have a red tail and yellow band to the abdomen, and nest underground or in trees.

Step into Nature

Common Carder Bee, *Bombus pascuorum*, (queen 13-17mm from April, worker 10mm, June), making a beeline for Stinking Hellebore, garden, April

A giant, velvety queen Red-tailed Bumblebee, *Bombus lapidarius*, (queen 17-20mm from May, worker 12mm, June), on Hellebore, garden, April

Queen Garden Bumblebee, *Bombus hortorum*, (queen 16-18mm, April, worker 13mm, May), on Aubrietia, garden, May

Queen Early Bumblebee, *Bombus pratorum*, (queen 13-16mm, April, worker 10mm, May), Clematis seed head, garden, May

Week 3

Encouraging biodiversity in your garden can be as simple as planting some seeds, Oxeye Daisy, Teasel, Honesty or Dame's Violet. Each plant can attract its own range of insects. Dame's Violet is a tall perennial with hairy stems and cruciform white, pink or lilac flowers. It is shaped to attract long-tongued insects such as hoverflies or bees, while the leaves (as part of the cabbage family) are very popular among the caterpillars of white butterflies.

But a special guest is a pale, perky micro-moth with long, forward facing antennae, *Plutella porrectella* generally associated with this, its larval foodplant. It is double brooded with larvae feeding during April-May and June-July.

Plutella porrectella, cocoon, garden, September

Plutella porrectella, (7-8.5mm, April to October), adult, garden, October

Dame's Violet, *Hesperis matronalis*, (90cm high, flowers 15-20mm, March to August)

It distorts the leaves with a small amount of silk, and then pupates in a delicate open-mesh, silk cocoon on the underside of a leaf. The moth flies during May and again in July-August and while predominantly nocturnal, can be found on Dame's Violet during the day. These garden finds in 2021 were the first time the moth was recorded in Offaly, and now we look forward to finding them every year. We will look for another member of this family, Diamond Back, *Plutella xylostella*, later in the year (June - W1).

A small selection of moths may be seen during the day but the majority of the 1,500 or so species in Ireland are nocturnal. These fly from dusk throughout the night as we are fast asleep, visiting plants for food or seeking a mate. Some plants like Honeysuckle are designed to be pollinated by moths and as such only start to release their fragrance at dusk. To catch sight of some of these incredibly varied species, we can avail of a moth trap. The bright light attracts moths who then fall down into the box and settle in the internal egg cartons to sleep. The trap and surrounding area is checked at dawn with much anticipation to see what delights are sleeping inside. Once identified and photographed, the moths are released back into surrounding shrubbery, away from hungry Robins.

March - Week 3

The shape, size, palette of colour and sheer numbers of moths found in the garden and local bog at Killaun each year is truly astounding. While only a few are found early in the year, the numbers and variety increase throughout the year to compulsive levels. Help with identifying moths is available from MothsIreland who also provide the required trapping licence. They can be recorded on MothsIreland or the National Biodiversity Data Centre (NBDC).

Some of the earliest moths to emerge are found in the garden and also frequent woodlands, generally flying until May, such as the variably coloured Common Quaker. It is one of a number from the same family, so named due to their sombre colours.

The brighter March Moth, with triangular shaped wings, is male - as the poor female is flightless and unable to leave the tree trunk. Mating shortly after emerging, she lays eggs on her own cocoon, dying shortly afterwards. There are a number of such unfortunate females among the many moth species. Hebrew Character is a frequent visitor, with a strong bracket-shaped mark mid-wing resembling the Hebrew letter *'nun'*, giving the moth its name. Early Grey with subtle markings, can also come in a soft, delicate pink.

Step into Nature

Common Quaker, *Orthosia cerasi*, (forewing length 13-17mm, March to May)

March Moth, *Alsophila aescularia*, (16-19mm, January to April)

Hebrew Character, *Orthosia gothica*, (15-17mm, March to June)

Early Grey, *Xylocampa areola*, (15-18mm, March to May)

March – Week 4

Week 4

We have had a garden pond for many years which brings marvellous variety on the surface of the water, within the water and to the surrounding area. Yet it was only last year, after sitting still for a few moments by the edge of the pond, that we noticed lots of little insects scuttling to and fro over Duckweed.

They are the wingless (usually) Water Cricket, with a striking red-dash on each side of their abdomen and inner white dots. A gregarious semi-aquatic bug, 5-7mm, with powerful legs for running over the water, it can be found in still and slow moving water and streams (although rarely in ponds). Adults feed on small insects that have fallen in the water. They lay their eggs in long rows on floating plants, in May and June, with nymphs appearing from June to August.

Water Cricket, *Velia caprai*, (5-7mm, March to September), garden pond

Lesser Diving Beetle, *Acilius sulcatus*, (15-18mm, March to October), garden pond

Also spotted by the edge of the pond enjoying some warming sunshine, was an endearing Lesser Diving Beetle, broad, oval-shaped, dark brown with a yellow tip to the abdomen and head. The female is distinguished by her hairy, 'go fast stripes' or furrows along the abdomen. She lays her eggs on land. Once startled, she turns swiftly, swimming down into the depths of the pond.

A diving beetle can live for several years, laying eggs each year. It has three pairs of legs, each with a specific function. The first short pair are for grasping hold of prey when eating, the middle legs are used to hold on to plants while resting, and the larger, rear legs are for thrusting through the water.

The grey-brown Large-grooved Diving Beetle with finely etched wing casings (elytra) along its back, is another welcome visitor to the garden pond.

Large-grooved Diving Beetle, *Colymbetes fuscus*, (15-20mm, March to November)

March - Week 4

A trip to a bog pond may provide broader interest and the local Killaun Bog is a veritable treasure trove. The Smooth Newt, Ireland's only native amphibian with a tail, lays individual eggs on water plants, folding over the leaf to protect it from predators, while allowing the eggs access to water. As the tadpoles hatch, they come together in a large tightly-packed group. Brown with hints of yellow and orange, the feathery gills are red-brown.

Tadpoles develop into small newts, front legs first. Their long tail allows them to glide quickly through the water, and slender fingers and toes are used to climb water plants in search of insects and worms. Maturity takes about 18 months, when they leave the water as 'efts' to live in damp moss, grass or Heather. They come back to water again as adults, in spring, to lay their eggs.

The graceful, 6-10cm adult has soft velvety skin, with a few dots along the female belly. Males have a low fringe along their back, stripes to the head and an orange belly with large dark spots.

Smooth Newt, female (top), tadpole-small newt (centre), male (bottom)

Step into Nature

Smooth Newt, *Triturus vulgaris*
at Killaun Bog

March – Week 4

Their appearance changes during mating season, which typically lasts from February to June, with spots becoming larger and hind feet more webbed. Males develop a high crest along their back and a light blue stripe along its tail as the belly becomes a rich orange. Not be left out of spring dress, the female then develops a low fringe along her back. Males perform a courtship dance and lead the selected partner over a sperm packet they deposited, which the female absorbs, thus fertilising the eggs internally. Newts are predominantly nocturnal and can live for up to six years.

A slender bug, the Water Measurer, found at the Cranberry Bog, Gallen (part of Noggusboy Bog), is much longer than the more common Pond Skaters, *Gerris* sp. Dark brown with restrained markings along the extended abdomen, their most distinguishing feature is the elongated head with bulging eyes located along its length. It moves slowly, raised off the surface of the water with long threadlike legs, keeping to the verges in search of food.

Water Measurer, *Hydrometra stagnorum*, (9-12mm, March to September), Cranberry Bog, Gallen and garden pond

Smooth Newt at Spring, female (top), male (centre)

Step into Nature

The boardwalk at Killaun Bog

March - Week 4 — Step into Nature

Also found at the bog, is the metallic Green Tiger Beetle with yellow spots, copper legs and antennae. It is usually seen (and heard) on a sunny day as a blur of green whizzing past, when it rises up from the ground ahead, then flies off to a gallingly distant location. It is especially striking against a backdrop of Reindeer Lichen. A common species along bog walks, it is a voracious predator of small insects, predominantly ants, by way of its sickle-shaped mandibles (jaws).

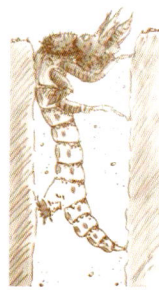

Deep, perpendicular tunnels in the ground contain their larvae who wait at the opening for passing prey, seizing them with hook-jaws. The remarkably unattractive larvae manoeuvre in the shaft by using a bristly point on their abdomen. After two years, they block the shaft and dig out a cell at the bottom where they transform into a pupa. Within a few weeks, it changes into beetle form, but remains below ground over winter, emerging the following year during warmer weather.

This Green Tiger Beetle, *Cicindela campestris*, (12-16mm, March to August), at the Cranberry Bog, Gallen, stayed still for a photograph as it was busily munching an ant.

March - Week 4

What a truly majestic creature with wings like folded chocolate and vanilla ice cream. This spectacular Puss Moth was resting on a tree by the moth trap, attracted to the light but far too regal to associate with other moths inside. A large white, fluffy moth with black spots around the thorax (between head and abdomen) and a stunning array of steep, wavy grey lines along the forewing with golden longitudinal strips. This was rather early as their normal flight times are May-July, when they visit gardens, hedgerows, woodland and bogs.

We will look ahead through their ever-changing life cycle to the somewhat startling larva seen July-September, feeding on poplars, Willows or Aspen (particularly saplings).

Shiny, red-brown, finely grained eggs are laid in pairs on the upper surface of a poplar or Willow leaf. Initially, the larva is a tiny black caterpillar with two long straight tails, which, when disturbed, extend red filaments. As the larva grows, becoming red then green in colour, by its final instar it has developed into the most extraordinary caterpillar.

Puss Moth, *Cerura vinula*, (larva July to September), early instar at Killaun Bog, July

Step into Nature

Puss Moth (29-38mm, March to July),
Cerura vinula, garden, March

March - Week 4

Puss Moth, *Cerura vinula,* final instar

Green with a long brown patch, trimmed with white along its back, it retains its two tails. When alarmed, the head can be drawn back into its body to the red ring next to it with a slit at the base from which acid can be squirted. Two dark spots look like staring eyes, which may deter hungry birds, while the lashing tail attempts to keep off its parasite, a large Ichneumon wasp, (see also August - W3).

Puss Moth, chrysalis

The caterpillars pupate in a red-brown chrysalis and overwinter in a hard cocoon of chewed bark and silk which is fixed to a tree trunk. The moth is assisted out of the hard cocoon due to clever planning by the caterpillar who leaves a thin layer at the exit end and a chrysalis case with a triangular shaped 'head'. The moth breaks the head end of the case, moistens the surrounding area, so it can force its way out of the cocoon, leaving the chrysalis case behind.

Puss Moth, cocoon

Possible Puss Moth cocoon with large exit hole, fixed to garden tree

Deep in old woods at the base of Hazel or Elm, in total shade, you may come across a rare, strange, colourless plant. As it is entirely parasitic, Toothwort does not need to produce chlorophyll so has no customary green colouring. Using underground rhizomes it takes nutrients from tree roots. The thick, flowering downy stems with multiple, milky-pink tubular flowers, look unnervingly like rows of molars.

Toothwort, *Lathraea squamaria*, (15-20cm high, April to May), Clara Esker Walk

On a brighter note, perhaps the most beautiful flower growing by hedgerows is Greater Stitchwort with large white flowers of five, deeply-cleft soft, ridged petals and a light green base. Weak, fragile stems gain strength by using pairs of long, stiff leaves against the adjoining grass, allowing the plant to creep higher through the bank. Beetles, flies, moths and butterflies visit the flowers for nectar and pollen. They were thought to relieve pains or stitches, giving its name.

Greater Stitchwort, *Stellaria holostea*, (up to 50cm high, flowers 15-30mm, April to June)

March – Week 4

On a sunny day towards the end of March, a native woodland walk is the place to be, to catch sight of early butterflies searching for food. The esker walk in Clara is a super location with a variety of habitats including woods and bog. We soon spotted a beautiful, large Brimstone butterfly, one of our earliest to appear, (we found one in Banagher at the end of February). Its lemon-yellow wings add a vibrant splash of colour for early spring, as only Hawthorn and Spindle are starting to show fresh lime-green leaves. They cover great distances with a strong, vigorous, flight that is a joy to behold. You might see a pair of Brimstones as they dance before you on a sunny morning, through woodlands, scrub and some lucky gardens. Described in *The Irish Butterfly Book*, by J. M. Harding, as the only Irish butterflies to live long enough to see their offspring fly. Eggs are laid singly on Buckthorn April-May, hatching some two weeks later with larvae munching entire leaves up to July.

Brimstone, *Gonepteryx rhamni*, (60-74mm, March to November), on Bluebell, garden, April

Step into Nature

Emerging butterflies feed voraciously into late summer, prior to hibernation.

Another flash of red and we see the stunning colours of the large Peacock butterfly with characteristic eyespots on all four wings. These butterflies overwinter from September to early spring and lay eggs in batches under Nettle leaves.

Anywhere there are flowers and Nettles, we will find the brightly marked, familiar Small Tortoiseshell from March to November. They are often found hibernating indoors during the winter and as they emerge in early spring, seek nectar from Dandelions and Willow.

This highlights the importance of allowing Dandelions and Nettles grow at road verges and in our gardens. Groups of eggs are laid on young Nettles, hatching in less than two weeks.

Small Tortoiseshell, *Aglais urticae*, (45-62mm, March to November), on Marjoram, August

Peacock, *Inachis io* (63-75mm, March to September), on Water Mint

March - Week 4

While on the esker walk at Clara, we were also fortunate to glimpse wings flashing orange, as a wonderful Comma landed, to bask in the warming sunshine. These butterflies have made a rapid progression across Ireland since they first arrived in 2000, lasting through the winter by 2014. We first found one in the garden in March 2022. The scalloped wings have a dark underside bearing the signature 'c-shaped' comma mark from which it is named. Hardy overwintering butterflies fly again from March to May and lay eggs singly on Nettle leaves. Within this short egg-laying timeframe, four spring butterflies - Peacock, Small Tortoiseshell, Red Admiral and the Comma - require Nettles in order to lay their eggs for the next generation. This is one small area where we can all help these beautiful, delicate insects, by allowing clumps of Nettles grow in our gardens.

On a balmy late-summer evening, a dancing pair were seen in the garden, gracing the sky with a mesmerising, elegant, twirling display overhead. Sweeping together from one end of the meadow to the other, yellow-buttered by the evening sun, one broke away to scan the spectator. I was standing still, soaking up the beauty of the moment, when the magnificent Comma flitted closer and closer, then came to land on my arm. The breathtaking surprise and exhilaration of the

Step into Nature

moment, was marginally tempered by wondering if a photograph was possible. Keeping the esteemed arm immobile, while trying to extract the phone from my back pocket, the nonchalant Comma was perfectly content to remain for numerous shots.

The palpable sense of thankfulness at witnessing such indescribable beauty so close is a sincere privilege and an unforgettable experience. Their heart aching beauty remains, bringing the summer back to me still.

Comma, *Polygonia c-album*, (50-64mm wingspan, March to August), garden, August

April

Week 1 Woodland walks are at their most pleasant during April, as many beautiful and prolific flowers spring forth. Most trees are still without leaves and the plants gratefully avail of increased sunlight, providing vibrant, fragrant carpets of Violets, Bluebells and Celandine. The lightly scented, graceful Wood Anemone flowers from the end of March in broad-leaved woodlands with slender-stalked palmate leaves (divided in five). A separate reddish stem with three leaves wraps around the fragile flower bud as it pushes up from the ground. The bud opens to show typically six white sepals (which have taken the place of petals), tinted with pink and a centre of golden stamens. The flowers visited by bees, beetles, flies and bugs, provided a vast range of former human uses. Flowers mixed with oil were said to assist hair growth; it was used to treat leprosy; and chewing the roots was said to assist with removal of phlegm.

Wood Anemone detail of sepals taking the place of petals

Step into Nature

Wood Anemone, *Anemone nemorosa* (20cm high, flower 2-4cm, March to May), at Birr Demesne

April - Week 1

Bank Vole, *Myodes glareolus*, (8-12cm long, c.18 month lifespan)

The rhizomes which provide a thick root system for Anemones are an important source of food for the Bank Vole, and it was at Cooleeny, with carpets of Wood Anemone, that we first came across one. Classed as an invasive species with medium impact, they have spread across Ireland from the south. Around 10cm in length, they look superficially like mice but are noticeably browner, have more rounded ears, a blunt nose and shorter tail.

As Wood Anemone finish flowering, we see a low, creeping plant with thin, red stems and yellow-green shamrock leaves. The lemon-flavoured leaves of Wood-sorrel close over with approaching darkness or storms. The wind-pollinated flowers droop gracefully with five white petals lightly tinged with purple. Showing remarkable resilience, another set of buds develop in the summer which, under the safety of leaf cover, grow into seed capsules, without ever flowering. When ripe, the long stalk rises, bringing the seed capsule up, allowing it to catapult the seeds away from the leaves below.

Wood-sorrel, *Oxalis acetosella*, (10cm high, flower 10-15mm, April to May)

Step into Nature

We can look forward to the uplifting sight of Primroses flowering in leaf litter, in woods and hedge banks, before the delicate Wood Anemones are blown. They have stored nutrients overwinter in their thick rootstock. The long, wrinkled, oval leaves grow in a circle around a central thick stem which houses hairy flower stalks. They are topped by the familiar pale-yellow flowers and completed with specks of darker yellow to the base of each petal.

Primrose, *Primula vulgaris* (20cm high, flowers 20-30mm, March to May), at Knockbarron

This contrasts with the butter-yellow of the taller Cowslip with central orange spots. They grace limestone hills, fields and road verges with juicy spring colours and a delicate fragrance. A drooping, one-sided umbel (cluster of flowers), on a short stalk is attached to the top of a tall scape (stem). Ointment made from the petals was thought to treat sunburn, a medicinal wine was made of the flowers, and leaves were boiled for eating.

Cowslip, *Primula veris*, (flowers 9-15mm, April to May), Tullamore road verge

April - Week 1

A hybrid of both, the False Oxlip occurs naturally when Primrose and Cowslip grow close together. They are a similar height to Primrose and the flowers are also similar, with five pale yellow petals. However, from their Cowslip parentage, they have an orange spot to the centre and are born from a umbel on a scape. Their individual flourish is the multi-sided cluster of flowers. Primrose comes first March-June, then False Oxlip April-May, followed by Cowslip May-June.

False Oxlip, *Primula vulgaris x veris = P. x polyantha*, (20cm high, flower 15-20mm, April to June), Mount St Joseph

Despite their tall, straight, dense floral spike, the Early Purple Orchid can be easily missed on the woodland floor. Sometimes it is the long, basal, brown-spotted leaves (yet not always spotted) that can be found first. The large, wide lower petal (labellum) of the complex flowers acts as a runway for visiting bees. As the bee moves up into the flower, sticky pollen-masses fix to its head, these later detach and stick to another Orchid flower.

Early Purple Orchid, *Orchis mascula*, (40cm high, flower 8-12mm, April to June), at Cooleeny

Common Twayblade, *Neottia ovata*, (50cm high, April to August), Cooleeny

Close by, another easily identified Orchid may be seen, the green Common Twayblade, (flowering up to August). It has two broad, oval basal leaves wrapped around the tall, slender spike, (up to half a metre). The numerous, petite, sophisticated yellow-green flowers are visited by small insects including bees.

Pineappleweed, *Matricaria discoidea*, (12cm high, flower 8-12mm, April to November), garden

Growing on bare or disturbed ground, Pineapple-weed appears like a form of 'petalless' Daisy with feathery leaves and a strong pineapple fragrance when crushed. It is a tenacious wildflower which found its way around Ireland by sticking to car tyres and the soles of shoes.

Around the edges of the lawn, or disturbed ground, you might find Field Madder, a dainty, low-growing, creeping plant with tiny, lilac cruciform flowers and small whorled leaves.

Field Madder, *Sherardia arvensis*, (flower 2-3mm, April to September), edge of garden lawn

April - Week 1

Garden moths at this time of year have to blend into bare, predominantly leafless branches and leaf litter, with their matching, subtle colour schemes. The Double-striped Pug comes from a group of similar brown or grey moths, which can cause identification problems. It may be recorded all year but is most common March to October, visiting parks, gardens, hedgerows and heathland. This is a fresh, strikingly marked example. They may be seen during hot sunny days but, more typically, fly at dusk to flowers including Holly, Ivy, Gorse, Broom and Heather among others, and overwinter as a pupa in plant debris.

Ypsolopha mucronella was an exciting discovery in 2023, a first for Offaly and only the sixth for Ireland.

Double-striped Pug, *Gymnoscelis rufifasciata*, (8-10mm, all year), garden

Ypsolopha mucronella, (13-16mm, August to June, adult hibernates), garden

Step into Nature

It is classed as a micro-moth - families of small, allegedly more primitive, species. The adult, though petite and fragile, survives the winter in Ireland, living in woodland and scrub and lays her eggs on Spindle.

The richly coloured Shoulder Stripe is an attractive, banded moth of purple-brown, cream and dark brown. Flying from March to May at dusk, it feeds on rose hips and lives in hedgerows, woodlands and gardens. Its larval foodplant is Dog-rose.

We found an Oak Beauty last year by a neighbour's front door and it was a first record for Offaly. This year, a male with impressive feathered antennae was found in the garden. When resting on an Oak tree, the patterned wings blend perfectly with branches covered with barnacled lichen.

Shoulder Stripe, *Earophila badiata*, (14-18mm, March to May), garden

Oak Beauty, *Biston strataria*, (17-27mm, late February to April), garden

April - Week 1

While out walking or even driving, the Yellowhammer, one of our small buntings, is very noticeable due to its striking, bright yellow plumage. It is a widespread species in Ireland, but unfortunately uncommon. They nest below hedges, feeding on insects, spiders, seeds, berries and fruit. We see them each year in a few neighbouring locations, close to arable farms, perched on top of the hedgerow or cooling down in roadside puddles.

The more common Siskin, with strong yellow and black markings, nests high in conifers and can be found in winter around Alder, Birch and Larch. We have been fortunate to have them visiting bird feeders in recent years, normally appearing only in the New Year.

Yellowhammer (top), *Emberiza citronella*, (wingspan 26cm), three year lifespan, eggs in May

Siskin, *Carduelis spinus*, (20-23cm wingspan)

Step into Nature

Week 2

It is a good time to take a closer look at south-facing, low earthen banks, by the road verge or field, where we have noticed solitary or mining bees. There are more than 100 species of bee in Ireland and 80 of them are solitary bees.

Sometimes you might be fortunate to notice tiny holes with a clump of discarded soil in the lawn as we did, directly under a swing or clothes line. Braving these somewhat hazardous locations, were numerous large, handsome Ashy Mining Bees. They are black with two furry, light grey thoracic bands. We may see them from March to June and as with most of our spring species, require Willow as a source of food, along with Gorse and Blackthorn.

Nest of Ashy Mining Bee on garden lawn

Ashy Mining Bee, *Andrena cineraria*, (13-16mm, March to June), a bit grumpy, garden lawn

April - Week 2

They are smaller and more slender than our bumblebees and do not live as workers with a queen bee. Each female solitary bee makes her own nest, leafcutters (*Megachile* bees, August - W2) with leaves or mining in earthen banks. She lays a fertilised egg with food of nectar and pollen. The larvae hibernate, emerging as adults the following year.

Orange-legged Furrow Bee, *Halictus rubicundus*, (9-12mm), garden bank

In a garden bank, we can see the Orange-legged Furrow Bee, medium-sized, with a black abdomen and fine white stripes, orange hind legs and females sporting a ginger-haired thorax (males, paler yellow with noticeably longer antennae). Females fly March to October (males mid-June to October) foraging on blossoming trees and a variety of flowers. While they have their own nests, they are eusocial (cooperative care of juveniles), mining in groups.

The Early Mining Bee or Orange-tailed Mining Bee, is a medium-sized bee.

Early Mining Bee, *Andrena haemorrhoa*, (11-13mm, March to August), garden

Females have a distinctive red-tipped, black abdomen, a rich ginger-red thorax and orange hind legs. They fly from March to August, foraging on Willow, Gorse, Hawthorn and Dandelions, where we find them in the garden, nesting singly or in loose groups.

Fresh Chocolate Mining Bee females have a red-brown thorax and black abdomen. Males are smaller, duller with long antennae. Flying from mid-March to July, they are particularly noticeable as Willows, Blackthorn and Hawthorn begin to blossom. They also visit a variety of flowers, umbellifers, Dandelions.

Nesting singly or in relaxed groups in leaf litter or on banks, it is often easier to locate the nest by the presence of their parasite bee, *Nomada marshamella* and sometimes *Nomada goodeniana*, for example.

Chocolate Mining Bee, *Andrena scotica*, (7-10mm, March to July), fresh and worn females, garden

April - Week 2

Nomada bees are slight, wasp-like bees which have devised a way of rearing their eggs without making their own nest. Instead they hover around the south-facing banks or leaf litter watching for the opening to their preferred host's nest, where they lay an egg. As the Nomada grub develops it destroys the host egg with its sharp mandibles (jaws) and as a kleptoparasite (taking food from another) eats their food store.

Nomada marshamella is a cuckoo bee which takes over nests of *Andrena scotica*. Females have a black abdomen with yellow stripes. They have two broken yellow stripes, differing from the similar *Nomada goodeniana* who have one, (frequently none). Another feature is the yellow (*N. goodeniana*), versus red spots on the thorax. Both females have legs and antennae that are orange-brown and usually two yellow spots on the thorax. *N. marshamella* flies from April to September, *N. goodeniana* flies from April to June and parasitises the Ashy Mining Bee. For this reason we have all four species in the garden feeding on a variety of flowers including Dandelion and Greater Stitchwort.

Nomada marshamella, male, two broken yellow stripes

Nomada goodeniana, one broken stripe, both species 9-13mm, garden

Step into Nature

By the fading light of day, we spotted this Nomada bee hanging on to a Birch leaf. It was fast asleep, fixed to the leaf by its mandibles, with all six legs dangling as it swayed rhythmically in the soft evening breeze.

Not so fortunate are the flies with enlarged, whitened abdomens, which have been parasitised by the *Entomophthora* fungus. The dead flies (or other insects) release spores which are ingested by more flies. This can result in a large localised group on the same plant (over twenty were found on long grass in the garden). The digested spores grow into the fungus internally, later breaking out again, as more spores.

A sleeping Nomada, with all legs dangling, held by mandibles alone

Flies parasitised by the *Entomophthora* fungus

April - Week 2

Bog Beacon is one of the earthtongue fungi, and while many grow in grass or leaf litter, this uncommon variety grows in bog streams, ditches or in Sphagnum Moss. It is gregarious (growing in groups) and it provides an essential service, breaking down plant remains. The vibrant yellow-orange head, circular or oval shaped, rises to no more than 4cm on a translucent stem. They appear as if by magic, like a tiny torch over the dark waters of the bog at Killaun.

Bog Beacon, *Mitrula paludosa*, (4cm, spring to early summer), at Killaun Bog

Step into Nature

Our largest Irish spider is the Raft Spider, shy, velvety, dark-brown, with two pale yellow longitudinal stripes on the side of the thorax and abdomen, and light spots scattered over the abdomen. It sits beside bog pools and lakes with its front legs on the water, surface feeling for the vibrations of potential prey moving through the water. It plunges into the water to grab them when they come near enough. It can catch and eat small fish and froglets.

A somewhat similar, paler spider, grey to orange-brown with longitudinal stripes, is the Nursery Web Spider. Both species carry several hundred eggs in a large white sack under their body and later guard the hatched spiderlings in a copious tent-like web. We have often spotted them soaking up the sunshine - Raft Spiders along the boardwalk at Killaun and Nurseryweb Spiders on large Cranesbill or Nettle leaves in the garden.

Raft Spider, *Dolomedes fimbriatus*, (9-20mm), at Killaun Bog

Nursery Web Spider, *Pisaura mirabilis*, (10-15mm), on garden Cranesbill, it is also common on open bog

April - Week 2

Lords-and-Ladies, *Arum maculatum*, (up to 25cm high, flower April to May), garden in the spring

Step into Nature

While walking through woods, along hedgerows or within the garden, we notice clumps of large spear-shaped leaves, known as Lords-and-Ladies. It is equipped with a bewildering pollinating strategy, inextricably linked with a tiny fly, the minute, furry, moth-like Owl Midges.

Owl Midge, *Psychodidae* sp. (4.5mm wingspan), garden

A central stalk rises up and unfurls as an elegant ellipsoid inflorescence. The outer hood is almost transparent, pale yellow-green, with a central pink to purple spadix (staff). These flowers have a distinctive, nasty smell, emitted around midday, which the Owl Midge finds extremely attractive, as it mimics cow-dung where they lay their eggs. The midge enters the outer hood, and starts its journey to the hidden chamber below. First it must push through band of hairs (sterile male flowers), then past the anthers (fertile male flowers), to reach the ripe pistil below; a group of both sterile and fertile female flowers (yellow carpels).

Lords-and-Ladies, *Arum maculatum*, chamber cross section, below flower

April - Week 2

The midge travels from one plant to another, carrying out the vital pollinating function. If this is the first plant the midge has been to and hence cannot shed pollen from another plant onto the pistils, the band of hairs provides an inflexible barrier, preventing the midge from leaving. Once pollen has been brought in, by another more adventurous midge, the stamens open, shedding their pollen on the midge. The hairs above then shrivel allowing any trapped midges escape, around two days later. While this may not seem too long, an adult midge only lives for a week.

Next autumn, we will see the resulting fruit (from the lower ring of female flowers), in the form of a bunch of bright red (poisonous) berries. In some locations, these are even more noticeable than the earlier flowers or leaves. One to watch out for from August to October and a source of food for our larger birds, thrushes, pigeons, blackbirds and pheasants.

The common name Lords-and-Ladies comes from the time of Queen Elizabeth I. The root was used as starch, to stiffen linen, and it allowed for the dramatic deep ruffs which then became popular, although presumably, highly uncomfortable.

Lords-and-Ladies, *Arum maculatum*, (berries up to 5cm, August to October), garden in autumn

Step into Nature

The starch was very abrasive and unpleasant to work with. In Ireland, in times of scarcity, there were attempts to diminish the acrid taste of the arum root to make it edible. First it was boiled, then baked, grated into water, and dried; thus providing a tasteless, allegedly nutritious powder.

As Easter approaches, we should look out for a low-growing flowering plant of the *Juncaceae* or rush family. There are three somewhat similar families, sedges, rushes and grasses. The old adage is most useful 'Sedges have edges, Rushes are round, Grasses have nodes from the top to the ground'.

Field Wood-rush or Good Friday grass (not particularly helpful including 'grass' in the common name), is found in fields or on lawns. It flowers from April to May.

The inflorescence are made up of three to six stemmed clusters of flowers and one separate stemless cluster which produce their seeds. They provide little splashes of colour in the lawn, with the creamy edges to the rich dark purple of the flowers. Due to their height, they usually go unnoticed, yet each year they stoically announce the forthcoming religious season.

Good Friday Grass, *Luzula campestris*, (10-25cm high, flower 2-4mm, April to May), garden

April - Week 3

Week 3

One of the most breathtaking things to see in nature or even better in the garden, is the different stages of an insect's life cycle. To come to know which plants they are likely to feed on as larva, and to become aware of the time of year that we may expect to see each stage. Witnessing the life cycle of the enchanting Orange-tip butterfly as it plays out amid the elegant Cuckooflower is a wonder to behold. Orange-tips are such a bright, beautiful, fresh butterfly that we hope to see flutter forth from April to June. Males are unmistakable with the vibrant orange end to their forewing, while both sexes have delicate green marbling under their hindwing.

The pattern allows perfect camouflage in dappled shade among the white flowers of another favourite plant Garlic Mustard, as it grows through hedge banks with Cow Parsley, *Anthriscus sylvestris*. They are found in damp bogs or woodlands and gardens where their foodplant grows, taking nectar for adults and for egg laying. When we discovered a small patch of Cuckooflower growing in the front garden, we let the area grow wild and cut it twice a year as a meadow.

Garlic Mustard, *Alliaria petiolata*, (up to 1m high, flower 3-5mm, April to June), Clonmacnoise

Step into Nature

A spiral pathway is mown through it to allow many pleasant hours meandering through, checking Dandelions for solitary bees or hoverflies and, as the year progresses, for butterflies. It was here that we first watched the beautiful Comma butterflies, noted in March - W4. But in April, we can find the Orange-tip, thanks to the ever expanding clusters of Cuckooflower.

Once spotted, we can look out for the tiny eggs laid singly on the Cuckooflower stem, just below the flowers, in full sunlight. The ovoid eggs are first white, turning yellow to orange over a week or two. The hatched caterpillar feeds off the flower pods, then the stalks (which helps us to find them) for up to three weeks. When ready to pupate, the larva selects a suitable plant stem (or wire fence) and, over two days, forms a green or brown pupa.

Summer spiral, Orange-tip, *Anthocharis cardamines*, egg, larva then pupa, garden

April - Week 3

Orange-tip, male, *Anthocharis cardamines*, (40-50mm wingspan, April June)

Orange-tip, female

Green-veined White, *Pieris napi*, (45-50mm, April to September)

This overwinters until the adult hatches the following April. Cuckooflower is connected by season with the familiar call of the spring Cuckoo. It is also known as Lady's Smock, as a moist field full of its flowers was thought to resemble linen laid out on a bleaching green. Initially, the four petaled flowers are a delicate lilac with leaves and stems resembling the cress family, and attempts were made to use leaves for salad.

Another butterfly favouring similar plants is the Green-veined White. It is best identified by its underwings, giving the butterfly its common name. So as April progresses with frustratingly cold, wet and windy weather, it is such a comfort to see the stalk of the Cuckooflower rising gracefully above long grass, knowing what is in store over the coming weeks.

Step into Nature

Cuckooflower, *Cardamine pratensis*, (40-60cm high, flower 12-20mm, April to June), in the summer spiral

April - Week 3

Our first blue butterfly of the year is the petite Holly Blue the only resident butterfly with a tree in its name. It is larger than the tiny Small Blue (16-25mm) and lacks the orange spots of the darker Common Blue (29-39mm, May - W2), both of which fly May to August. Flying high around bushes and trees, it can be hard to catch more than a glimpse of Holly Blues, unless it is a sunny day when they may rest with open wings on Holly or Ivy. They may be found in hedgerows, woodland and gardens. The underside of their wings are pale blue-grey with fine black spotting. Other foodplants include Alder Buckthorn, Gorse and Dogwood. Mint green eggs are laid singly on Holly or Ivy flowers depending on the time of year, as there may be two or even three generations in one year.

For further detail, refer to *The Irish Butterfly Book*, by J. M. Harding. There are fascinating discoveries about interactions between ants and blue butterflies. Ants protect and 'tend' the pupa for its sugary secretions and are thought to be attracted by the 'song' produced by the pupa.

Holly Blue, *Celastrina argiolus*, (c.35mm wingspan, April to May and July to September), on garden Euphorbia

Step into Nature

Unfortunately, we cannot hear this mysterious pupa song, but are fortunate to have the lovely Holly Blue visit the garden, typically in April.

In comparison, the fast flying Red Admiral appears huge with its glossy black wings and rich orange-red and bright white markings. The underside of the wings has a more restrained pattern, reminiscent of ancient charred parchment. It takes time to rest and feed, providing a tranquil sight in the garden. The Red Admiral is a migrant which visits Ireland each year, flying from southern Europe and North Africa and may be seen throughout the year, (more typically April to October).

Red Admiral, *Vanessa atalanta*, (64-78mm, April to October), on garden Oxeye Daisy, August

Red Admiral, *Vanessa atalanta*, underside of wings, on Field Scabious, Turraun Bog, August

April - Week 3

There are many unusual plants that are specific to raised or blanket bogs. The Midlands are fortunate to have several bogs and one such flower was selected as the county flower for Offaly. Bog Rosemary is a sparse, slender plant which can be difficult to find unless the graceful, pendant pink flowers are in bloom. They peep above the Sphagnum Moss at Clara, Killaun and Clonbeale.

Bog Rosemary, *Andromeda polifolia*, (40cm high, flower 8-10mm, April to June)

Across the bog or on damp mountains, we find a low shrub with angular, green stems. Bilberry, or 'Frochan' in Irish, has small oval leaves and attractive, solitary, pendulous orange-red flowers. Large areas of the Slieve Blooms are cloaked with Bilberry and if we seek them out again late summer, we will find a small black fruit, as they are part of the same plant family as Blueberries. These were traditionally picked on Frochan Sunday, the last Sunday in July.

Bilberry, *Vaccinium myrtillus*, (60cm high, flower 4-6mm, April to July), in flower at Clara

Bilberry, *Vaccinium myrtillus*, (berries 5-10mm, July)

Step into Nature

Round-leaved Sundew, *Drosera rotundifolia*, (20cm high, flower 5mm, June to August), Clooneen Bog, July

Round-leaved Sundew, *Drosera rotundifolia*, (leaves 1cm across), Killaun Bog

Great Sundew, *Drosera anglica*, (15cm high, June to August), Clara Boardwalk

Sundew is a small low-growing insectivorous plant of bogs and damp, peaty places. Thin red stems hold a rosette of different shaped leaves, depending on the species. They are densely covered with long red hairs, each tipped with a sticky fluid giving the appearance of a heavy morning dew. It is the leaves which sustain the plant as they develop from springtime. When an insect lands on the leaf, it remains caught in the gluey substance as first the hairs, then the entire leaf fold inwards, suffocating the prey. An extraordinary statement in the *Wildflowers of Offaly*, by John Feehan, states that Great Sundew trapped over six million Small White butterflies on a hectare of bog.

In July, slender stems bear a few petite white flowers, each with five petals.

April - Week 3

The opening of these flowers was the cause of great discussion as some didn't open at all, but as they are self pollinating, opening is not essential. The three different species are defined by their common name, the more frequent bright red, Round-leaved Sundew grows in many Offaly bogs, both on exposed peat and nestled among sphagnum. Dark red Oblong-leaved Sundew has 3-10cm long leaves and is found in wetter areas of the bog. Flowers grow from the side of the basal rosette (from the centre in the other two species). Great Sundew, as its name would suggest, has much longer upright leaves, green to the front, red to rear. We need to look in bog drains to find this impressive species, such as those at Clara boardwalk.

It is time to leave the bog and take a walk through the local wood again, where we may find some of the umbellifers or Parsley family in flower. They are plants with a cluster of flowers shaped like a parasol and come from the Latin, umbrella meaning 'little shade'. We can find many varieties throughout the year and one of the most delicate is Pignut. It is shorter than its similar, more prolific, relatives Cow Parsley or Hogweed, and is generally found in old fields or woods, rather than hedgerows. Carrot-like leaves rise from a basal rosette, while the 'flowers' are a combination of 10-20 umbels (flower cluster) which are each made up of 10-20 individual flowers. The common name comes from the edible nut-shaped root, highly favoured in earlier days by free-roaming pigs.

The taller, more vigorous, Cow Parsley is the first to fill hedgebanks with a frothy profusion of filigree flowers. The flowers are visited by a vast range of insects: flies, beetles, bees, wasps, bugs and butterflies and hedgerows are a buzz of activity from April to June. Another interesting visitor to look out for may be found by examining the leaves. We need to find a leaf with a white-brown tip, which may be home to a leaf-mining fly *Phytomyza chaerophylli*. Larvae mine the upper surface of the leaves and, when mature, make a slit in the lower surface to exit and pupate in the soil below. The hollow stems are used by insects during the winter.

Pignut, *Conopodium majus*, (50cm high, umbel 6cm, April to June), old woodland, Ballykealy

Cow Parsley, *Anthriscus sylvestris*, (1m high, April to June)

Cow Parsley leaf with *Phytomyza chaerophylli*, fly leafminer at tip

April - Week 4

Bluebells, *Hyacinthoides non-scripta*,
(40cm high, flower 14-20mm across,
April to May), at Charleville Demesne

Step into Nature

April - Week 4

Week 4

As Lesser Celandine continues flowering in large, vibrant swathes and Bluebells are starting to peep out, we also hope to find Wild Garlic or Ramsons. It is a vigorous plant growing in hedgebanks and across damp woodland floors. The pungent smell of garlic will announce their location before we even see the long, elliptical, basal leaves or the bright white, star-shaped flowers.

Ramsons, *Allium ursinum*, (50cm high, flower 15–20mm, April to May), with Bluebells at Charleville Demesne

Step into Nature

Leaves add flavour to salad and make a wonderful pesto. Highly esteemed in the past for medicinal purposes, it was also said to flavour cow's milk and meat, where eaten by livestock in fields. They are a nectar rich flower, much frequented by bees.

But it is always worth a closer look at the bees as one may, in fact, be a hoverfly. There are 180 species of hoverfly or syrphids in Ireland and many mimic bees or wasps. They feed on nectar and pollen, particularly yellow and white flowers. We will first look at the Ramsons Hoverfly, *Portevinia maculata*, which is confined to woodlands and old hedgerows where Wild Garlic grows. Short, stubby, orange antennae and an abdomen with three wide grey bands, (two broken), allows us to identify this hoverfly with some confidence.

Hoverflies are from the Diptera or true flies family, they only have two wings unlike bees and wasps who have four. They also have short, stubby antennae and very large eyes that wrap around the head ('Hoverflies have big eyes, bees have long antennae').

Ramsons Hoverfly, *Portevinia maculata*, (wing length 6-8.25mm, late April to May)

April - Week 4

But how can we tell bees and hoverflies apart when we see them out in the field? Bees fly with great purpose, zipping from flower to flower, while hoverflies, as their name would suggest, often hover mid-air. As they hover, checking the area for nectar or predators, it is a great opportunity to catch a glimpse of their abdomen (for identification) as their wings beat so fast they become practically invisible. We usually only notice the adults who live for a few days or weeks depending on the species. As with most insects, the adults seek nectar, pollen and a mate to produce eggs. Long, oval eggs hatch in less than five days and the short, flat larvae have three instars (or changes). The final instar lives for weeks to years, feeding on aphids and insect larvae.

One of the most common and straightforward hoverflies to identify is *Rhingia campestris,* which flies from the end of April to September, over two generations. Known as the Common Snout Hoverfly, it is the only one to have a long snout, and the orange abdomen is highly visible when flying.

While most hoverflies derive their nectar from yellow and white flowers, *Rhingia campestris* (because of a long nose), can successfully visit flowers with deeper tubes, like Bluebell and Red Campion.

Eristalis pertinax belongs to a widespread group with an interesting wing pattern which separates them from most other species. It has a downward 'loop' near the outer edge or the wing. *Eristalis pertinax* may be identified by its yellow tarsi (last third of their leg, equivalent to a foot) and triangular shaped abdomen.

Helophilus sp. are another group with the same wing loop, but are quite different in appearance. Similar in size, they have very striking markings, a black thorax with vertical yellow stripes and also sport a black and yellow-orange abdomen.

Male *Helophilus pendulus* are easier to differentiate and may be found in a broad range of habitats, including gardens.

Rhingia campestris, (wing length 6-9mm, April to September), female, garden

Eristalis pertinax, (8-12mm, February to November), male, Kinnitty Demesne

Helophilus pendulus, (8-11mm, April to September), male, Knockbarron

April - Week 4

Epistrophe eligans, (6-9mm, April to May), female, garden

Syritta pipiens, (4-7mm, April to August), female, garden

Eupeodes luniger, (6-10mm, April to September), female, Cranberry Bog, Gallen

A particular favourite for this time of year is the gleaming *Epistrophe eligans*. Good numbers of these rest on south-facing leaves in the garden, with their brassy thorax, glossy yellow and black abdomen glowing in the sunshine.

Another garden variety is *Syritta pipiens*, a smaller, slimmer hoverfly with impressively defined legs, which, happily, cannot be confused with other Irish species.

On a cold, bright sunny day at the Cranberry Bog, we find the air filled with the coconut fragrance of blossoming Gorse. Where sheltered from wind and facing into the sun, numerous insects are buzzing around including *Eupeodes luniger*. It is a well-marked, yellow-and-black hoverfly whose larvae feed on aphids.

Step into Nature

While admiring the hoverflies, we were surprised to find a beautiful Large Red Damselfly, resting on Gorse, while waiting for the sun to come back out. Usually, from May to September, with a whisper of wings, we can hope to see a mix of Damselflies and Dragonflies, so it was quite a treat to see one so early. Damselflies are smaller and more fragile than the sturdy, larger dragonflies. This is the easiest to identify as there is only one red-bodied damselfly in Ireland. It is widespread and may be found around bog pools, small lakes, streams (and, fortunately, the garden hedgerow), flying from April to August.

Large Red Damselfly, *Pyrrhosoma nymphula*, (c.33mm long, April to August), Cranberry Bog, Gallen

April - Week 4

Pine Ladybird, *Exochomus quadripustulatus*, (3-4mm, all year), Cranberry Bog, Gallen

Heather Ladybird, *Chilocorus bipustulatus*, (3-4mm, all year), Cranberry Bog, Gallen

On the same Gorse bush at the Cranberry Bog, Gallen, while waiting for the hoverfly to settle, we were delighted to see a Pine Ladybird, first recorded in Ireland in 2014. They are predominantly shiny black, small ladybirds, with four red spots; two to the back of their elytra (hardened forewings) and two hooked spots towards the front. There are only 10 recorded locations (to date), across Ireland. Within Offaly, we have found them several times at the Cranberry and Turraun Bog and at Cloghan Lakes.

We have found the even more elusive Heather Ladybird twice in heathlands (as their name would suggest), though not in the Heather but high on a Birch tree. On another occasion one was resting on rushes directly over Pallas Lake, a most unusual location.

Step into Nature

Orange Ladybird, *Halyzia sedecimguttata*, (4-6mm, all year), Cranberry Bog, Gallen

Cream-spot Ladybird, *Calvia quattuordecimguttata*, (4-5mm), garden

The Heather Ladybird is the same size as the Pine Ladybird, black with six red spots in a row towards the front of the elytra. Another form has a single spot on either side and two fused close to the centre.

The Orange Ladybird, a personal favourite, feeds on mildews in grasslands, woods and, after much searching, was found in the garden. It is orange with 14 white spots and has a charming transparent edge to the elytra and pronotum (a protective shield over the thorax, behind the head), like a transparent visor.

We have come across Cream-spot ladybirds in woods, at Turraun and Clooneen Bog, and in the garden. It has 14 cream spots on a reddish-brown background. Most ladybirds feed on aphids, assisting gardeners.

April - Week 4

Two-banded Longhorn Beetle, *Rhagium (Hagrium) bifasciatum*, (12-22mm, March to July), Cranberry Bog, Gallen

Step into Nature

Is this not the most magnificent creature? An imposing Two-banded Longhorn Beetle gripping the tip of a Birch branch, which was swaying precariously in high winds at the Cranberry Bog. It lays eggs in decaying pine trees, over two years, larvae bore deep tunnels until ready to pupate.

Before we move into May, we will have a look at a large clump of bright yellow Marsh Marigold. It is found in damp areas, by streams, bog pools and wet grassland. The petalless flowers are composed of (usually) five sepals and the shiny leaves elongate when it has finished flowering. Highly valued, these were used to 'cure' a horse bite, treat anaemia, epilepsy, warts and, if worn, could ward off unkind remarks. Within the flowers, clusters of the tiny, fetching, day-flying micro-moth, *Micropterix calthella* gather to feed on pollen (also on Buttercups and sedge flowers). They have a tufty yellow head and bronze-golden forewings with a purple base.

Marsh Marigold, *Caltha palustris*, (30-60cm high, flower 10-50mm, April to August), by the canal, Daingean

Marsh Marigold, with micro-moth *Micropterix calthella*, (3.5-5mm, April to July), Kinnitty Demesne

May

Week 1   May is such a wonderful month, full of promise, new life and fresh growth. Birch trees are delicately enfolded with soft, light green leaves, the crinkly leaves of Hazel are coming forth and Rowan trees are starting to blossom.
What of our well-known mammals? The much-loved Hedgehogs with their impenetrable coating of 6,000 spines (initially soft, stiffen as they mature) on a large body with short legs, neck and head. Sleeping by day, at night, they forage for beetles, worms, and slugs in gardens or parks. They hibernate from November to March in a leaf or moss-lined nest built under leafy mounds or undergrowth. Hedgehogs can swim and are good climbers, they mate for life producing four to seven young before September.

For some reason, the Red Squirrel always appears to be mischievous as it runs up the tree, around the trunk or along branches: it peeps down at us with such large, bright, black eyes and grips the tree with hairy soles and sharp, curved claws, and builds dreys (nests) in the tree for resting and for future offspring.

Step into Nature

Hedgehog, *Erinaceus europaeus*, (20-30cm long), at Killaun Bog

Red Squirrel, *Sciurus vulgaris*, (25cm long), Birr Demesne

Rabbit, *Oryctolagus cuniculus*, (30-40cm long), Ballinagar

May - Week 1

The larger offspring drey is spherical with a side entrance, sometimes adapted from a crow's nest using twigs, bark, moss and leaves.

Young Rabbits, known as a kits, develop strong back legs like Hares, allowing them to run at speed. The soles of their feet are covered with a thick, protective layer of hairs. Three types of hair cover their back, a dense, woolly under layer, strong coloured hairs, and another layer of longer hairs, all of which thicken up over the winter. Litters vary in size from two to eight and young are almost independent after just a month.

Walks are now enhanced with the distinctive call of the Cuckoo visiting Ireland from Africa until July. They lay eggs in the nests of other birds. Their young remove all other eggs or chicks and the often much smaller foster parent is left to feed the vast Cuckoo chick.

While walking through woods or bogs, we should look out for the great variety of caterpillars, feeding on their preferred plant (Birch leaves, Heather, Broom). The following are examples of some of those found and the adult moth. There is an extraordinary change from caterpillar to adult, with incredible efforts at camouflage depending on the favoured food source. Yet the Emperor caterpillar is bright green with pink spots, where could it possibly hope to hide?

Step into Nature

Emperor Moth larva, *Saturnia pavonia*, perfectly camouflaged on Ling Heather at Killaun Bog, June and adult (27-41mm), Killaun Bog, May

Pale Tussock larva, *Calliteara pudibunda*, Finnamore Lakes, September and adult (21-31mm), garden, May

Dark Tussock, *Dicallomera fascelina*, caterpillar, Killaun Bog, May

Dark Tussock (18-28mm), adult, garden, June

Garden Tiger larva (or caterpillar), *Arctia caja*, Cranberry Bog, Gallen, May and adult (28-37mm, July to August), July

Pebble Hook-tip larva, *Drepana falcataria*, on Birch at Cranberry Bog, September and adult (17-21mm, June to April to September, two generations), Killaun Bog, June

Scalloped Hook-tip, *Falcaria lacertinaria*, larva on Birch at Killaun Bog, July

Scalloped Hook-tip (14-18mm, April to August), Killaun Bog, May

Step into Nature

Oak Eggar larva, *Lasiocampa quercus*, on Gorse at Cranberry Bog, Gallen, May and adult (25-40mm, July to August), Killaun Bog, July

Pebble Prominent larva, *Notodonta ziczac*, Killaun Bog, July and adult (17-24mm, April to August, two generations), garden, May

Grass Emerald, *Pseudoterpna pruinata*, larva on Broom at Killaun Bog, May

Grass Emerald (14-19mm, June to August), Killaun Bog, July

The mesmerising Emperor Moth, *Saturnia pavonia*, (27-41mm, April to May), Killaun Bog, May

May – Week 1

Across our bogs there are some very unusual plants, particularly associated with damp, peaty conditions. One such plant is Bogbean, whose large leaves rise up from the water on slender stalks. Five white petals with pink-tinged margins are covered with an extraordinary silky, white fringe, giving it a truly unique appearance. Found in bog pools or large congregations as at Lough Roe, Gloster, their interlacing roots help to bind the soft soil in such locations. They were used as a remedy for rheumatism.

The minute Bog Stitchwort is more difficult to find but well worth the effort. We found the beautiful, pure white, starry flowers in loose clusters, on wet ground, when tying a shoelace. It is the same family as Chickweed and Greater Stitchwort, with five, small petals split to the base over the longer, pointed, green sepals. Common Butterwort, like sundew, is another insectivorous bog plant. Slimy, rolled yellow-green leaves form a basal rosette which traps insects.

Bogbean, *Menyanthes trifoliate*, (15cm high, flower 15mm, March to June), Finnamore Lake
Bog Stitchwort, *Stellaria alsine*, (25cm high, flower 5-6mm, May to June), Ballykealy
Common Butterwort, *Pinguicula vulgaris*, (15cm high, flower 10-12mm, May to June), Cranberry Bog

Their enzymes then digest and finally absorb their prey. Graceful stems hold a single hairy purple flower with a white centre and a long spur to the rear.

One of our most prolific and attractive wildflowers is Forget-me-not. With a profusion of petite blue flowers and bright yellow centres, it grows in waste places. It is a wonderful addition to a garden and popular with butterflies. Another plant of gardens, hedgebanks, verges and much visited by bumblebees, is Bush Vetch. A tenuous scrambling plant with slender, winding stems, oblong leaflets and pairs of pale purple flowers. It climbs up a firmer neighbour by way of clasping tendrils. Common Vetch (same family, yet not so common), is found in grassy places. It has brighter pink, pea-type flowers.

A plant that brightens our woodland walks with tiny, ice-white cruciform flowers is Woodruff. Flowers rise from whorls of pointy, glossy leaves which smell of hay and vanilla.

Field Forget-me-not, *Myosotis arvensis*, (25cm high, flower 3-4mm, April to June). Bush Vetch, *Vicia sepium*, (1m high, flower 15mm, April to October). Common Vetch, *Vicia sativa*, (80cm high, flower 18-30mm, April to September). Woodruff, *Galium odoratum*, (25cm high, flower 3-5mm, May to June)

May - Week 1

Flower Crab Spider, *Misumena vatia*, female, (10mm), with prey, white and yellow version

Flower Crab Spider, male, (4mm), black and white

Step into Nature

A small and highly successful hunter within garden flowers is the Flower Crab Spider. It lies in wait with outstretched front legs and uses venom (harmless to humans) to immobilise much larger prey. The female is more than twice the size of the black and white male, and can change colour depending on the flower that it rests on. A complete change from white to yellow, may take 25 days.

One of our most attractive flies in the garden or close to fields with cattle, and one of the few to have a common name, is the Noon Fly. Often found sunbathing on field gates or tree trunks, its glossy black thorax and abdomen acts as a wonderful backdrop to the characteristic wings with orange bases. Large, gold-rimmed eyes give it a slightly roguish appearance. Conopids, medium-sized 'robber' flies, are wasp or bee mimics. They wait on flowers until a bee lands, hop on its back and oviposit into its underbelly, parasitising it.

Noon Fly, *Mesembrina meridiana*, (9-13mm long, April to October), garden

Conopid flies, *Sicus ferrugineus*, (8-13mm, May to September), Killaun Bog and *Myopa* sp., garden

May - Week 1

Current solitary bees include *Andrena praecox*, an early, furry mining bee; males have a small, characteristic projection under their mandibles. They favour Willow-rich habitats and mine in a range of south-facing, light soil types including clay. This was a first record of the species in Offaly. *Andrena wilkella* favours legumes such as Bird's-foot-trefoil, Vetch and Clover, also mines in a range of soil types including heavy clays.

Blood Bees are dramatic in both name and appearance, with a deep red and black, shiny abdomen. They are kleptoparasites of ground-nesting bees including Orange-legged Furrow Bees, (April - W2) and we are most likely to find them at south-facing banks. Females enter the host nest, open a cell, destroy the resident egg or grub, then lay their own egg and reseal the cell. Most require microscope photography for species identification.

Andrena praecox, (7-9mm, February to May), and *Andrena wilkella*, (7-8mm, April to July), garden

Blood Bees, *Sphecodes* sp., (c.4-6mm, April to September), Clonoghill, Birr

Step into Nature

Female cuckoo bumblebees have no pollen baskets and are more shiny than social bumblebees. Female Four Coloured (or Forest) Cuckoo Bees, are very fluffy with a noticeable downward curve to their abdomen. They visit spring and summer flowers including Dandelion, thistles and vetches. They are a social parasite of the Early Bumblebee (March - W2) and *Bombus jonellus*, exploiting the host colony to rear their own brood.

Four Coloured (Forest) Cuckoo Bee, *Bombus sylvestris*, (13-15mm, March to September), garden

May - Week 2

The boundless beauty of hedgerows overflowing with Hawthorn, Ridge Road, Birr

Step into Nature

Hawthorn, *Crataegus monogyna*, (up to 6m, flower 12mm, May to June)

May – Week 2

Small White, *Pieris rapae*, (c.50mm, April to October)

Wood White, *Leptidea* sp., (c.42mm, April to July), garden

Dingy Skipper, *Erynnis tages*, c.39mm, May to June), Cranberry Bog, Gallen

Week 2 Time to look at some of the glorious butterflies now appearing. First, the Small White with dark wing tips, broader towards upper side of dull white wings. The female has two black spots, male one. Host plants include Nasturtium, Hedge Mustard and Charlock. The Wood White has narrow wings with rounded tips. The wings have an off-white upperside with black tips, while the underside is dusty white with light green and grey. Found in sheltered grassy areas with plenty of flowers, it rests with long, oval wings closed and has a more languid flight than other whites.

Brown and grey with white markings, the Dingy Skipper looks like a moth but is in fact a butterfly. They have clubbed antennae and occur in sunny, open sites on exposed ground with Bird's-foot-trefoil.

Step into Nature

The Small Copper is a lively orange and brown butterfly with dusty, grey hindwings. There can be no mistaking the dazzling Green Hairstreak as it lands with wings closed on yellow Gorse, showing its iridescent emerald green underside. It can be surprisingly difficult to see in flight due to its small size and dark brown, upper surface wing colour.

Common Blues are a picture of elegance. Males have exquisite blue wings fringed with white. Females are distinguished by their outer margin of orange and black. Both have a stunning underside of grey-brown with white-ringed black dots and orange markings to the outer margin.

Small Copper, *Lycaena phlaeas*, (c.33mm, April to October)

Green Hairstreak, *Callophrys rubi*, (c.33mm, April-July)

Common Blue, *Polyommatus Icarus*, (29-38mm, April to May), male (female centre), Cranberry Bog, Gallen

May - Week 2

Bird's-foot-trefoil is an important larval foodplant of many May butterflies with rich, yellow pea-type flowers that brighten scrubland, meadows and gravel areas. Hedge banks and woodlands are enlivened with the slightly scraggly, yellow flowers of Wood Avens (or Herb Bennet). The root was dried and stored with clothes, for its clove-like scent, to ward off moths. Another *Geum* species, Water Avens, is an impossibly elegant plant of damp ground or bogs. The graceful, drooping flowers have muted apricot petals encased in dark red sepals, a charming combination. A shorter plant of boglands or peaty grassland is Lousewort, which is semi-parasitic on the roots of grasses. Small spikes rise above crinkled, fern-like leaves, bearing pink flowers.

Yellow Pimpernel is a cheery plant with evergreen leaves that creeps along the woodland floor and damp, shady places. The five-pointed, satin-yellow, overlapping petals are fused at the base.

Bird's-foot-trefoil, *Lotus corniculatus*, (10cm high, flower 15mm, June to August). Wood Avens, *Geum urbanum*, (60cm high, flower c.15mm, May to August). Lousewort, *Pedicularis sylvatica*, (20cm high, flower 20mm, April to July). Yellow Pimpernel, *Lysimachia nemorum*, (40cm, flower 10-15mm, May to August)

Step into Nature

A wild hybrid Water Avens, *Geum x intermedium*, (50cm high, flower c.10mm, May to September)

Wild Garlic, or Ransoms, *Allium ursinum*, (35cm high, flower 15-20mm, April to May)

May - Week 2

A few years ago, while walking in Killaun Bog, we heard a loud whirring, like a dragonfly in flight. Suddenly, a large insect landed low on the trunk of a Birch tree, right beside the boardwalk. It was a large red and black type of cranefly. The shiny, seldom recorded impressive *Tanyptera atrata*, is a female with a sabre-like ovipositor to deposit eggs in soil.

Tanyptera atrata, (15-19mm, May to June), at Killaun Bog

When passing Birch trees, always look out for leaves rolled in a tight cigar shape, common on 10-20-year-old Birches. They are made by the female Birch Leaf Roller. She makes S-shaped incisions on the leaf allowing it roll in a complex manner, to both protect and feed her young, pale and legless larvae, which later pupate in the ground.

The adult weevil has a straight snout and thick rear thighs. Due to the precise incisions and intricate final shape, the rolls are of great mathematical interest.

Birch Leaf Roller, *Deporaus betulae*, leaf tightly rolled, Killaun Bog

Birch Leaf Roller, *Deporaus betulae*, (3-5mm, May to August), adult, garden

May - Week 2

Euura pavida, adult female (6-7mm, May-June + July- Sept), checking for suitable leaf, garden, May

Euura pavida, black thorax and yellow-orange abdomen and yellow legs

Euura pavida, adult female ovipositing on underside of Willow leaf May, (also on poplars and Alder)

Euura pavida, leaf turned to view shinny, oval eggs laid in neat rows over three leaves, garden, May. Larvae hatched 12 days later

Euura pavida, close up of well-developed single larva

Euura pavida, typical gregarious feeding group on Willow

Step into Nature

It is a good time to take a look at some of the many Sawflies (Symphyta) that we can encounter in the garden, woodlands or bogs. They are a broad-ranging, fascinating group with numerous species in Ireland. Many adults may be encountered casually as either adults or their larva, munching on leaves. They vary from solitary feeders such as the large *Cimbex femoratus* larva which feeds on Birch, to the smaller gregarious *Euura pavida*. Some of the adults are very similar, making identification frustratingly difficult or impossible in the field, so their larval food source can of great help in this process. It is often a more convenient way of securing a firm identification. The distinctive (at this instar), large *Cimbex femoratus* larva was first encountered at the Cranberry Bog, Gallen, in September while the vast, magnificent adult rested on a sagging Birch leaf at Clooneen Bog in May.

Cimbex femoratus, larva on a Birch leaf Cranberry Bog, October

Cimbex femoratus, (20-28mm, May to August), adult on Birch, Clooneen Bog, May

May - Week 2

Common Blue Damselfly larva, *Enallagma cyathigerum*, final instar

Four-spotted Chaser larva, final instar

Chaser-type exuvia at Lough Boora

Dragonflies and Damselflies come from the insect order Odonata. The life cycle begins with eggs, laid late summer into plant tissue, soil or scattered over the water, depending on the species. They hatch into larvae, called nymphs, which moult through six and 18 instars (stages). As with the adults, dragonfly and damselfly larvae differ in appearance. At this stage, they are already predators, catching prey underwater. When ready to metamorphose, they climb up the stems of water plants.

The fully-formed adult waits inside this larval skin. It can take up to three hours for the dragonfly to fully shed its exuvia (larval skin). They then make their first tentative flight, in large numbers from sheltered lakes, ponds or streams. One of the first dragonflies to emerge, in great numbers, is the Four-spotted Chaser.

Step into Nature

Four-spotted Chaser, (39-48mm, May to August), note dark wing base and spot on each wing, giving its name. Found at lowland lakes and ponds

Hairy Hawker, (55mm, May to July), with hairy thorax and tear-drop spots on abdomen, the only hawker flying before mid-June at fens, lakes, pools and streams

Four-spotted Chaser, *Libellula quadrimaculata*, *June*. What a treat as it zoomed around the garden then landed for a photograph!

Hairy Hawker, *Brachytron pratense*, sheltering in Gorse at Cranberry Bog, Gallen

May - Week 2

Common Blue Damselfly, *Enallagma cyathigerum*, (32mm, May to August), Club-shape

Variable Damselfly, *Coenagrion pulchellum*, (33mm, May to August), Wine glass

Azure Damselfly, *Coenagrion puella*, (32mm, May to August), U-shape

Blue-tailed Damselfly, *Ischnura elegans*, (31mm, May to September), Blue tail

Three similar blue and black damselflies are best identified by checking the second abdominal section, below the thorax. The Common Blue Damselfly has a distinguishing club-shaped mark below the thorax. It also has thick blue wedges on the thorax but no black spur (shorter black mark) to the side.

The distinctive mark of Variable Damselflies is in the shape of a wine glass. They do have a black spur to the side of the thorax.

Azure Damselflies have a U-shaped mark (male), and also have a black spur to the side of the thorax.

The remaining Blue-tailed Damselfly has a black abdomen and, as its name would suggest, a blue tail, one of the few identifiable in the air. It is also a good time of year to see some of the vibrant day-flying moths on the bog.

Step into Nature

Narrow-bordered Bee Hawk-moth, Hemaris tityus, (18-21mm, May to June), Cranberry Bog, seen as a blur of wings dashing from one flower to another, on Bird's-foot-trefoil. Clonmacnoise, a mating pair.

Common Purple and Gold, *Pyrausta purpuralis*, (7-11mm, March to September), Clooneen Bog. A micro-moth (yet Small Purple-barred is a macro-moth), flying in sunshine in grasslands; chalk, limestone and coastal. Their larval foodplant, Thyme, must also be nearby

Small Purple-barred, *Phytometra viridaria*, (9-11mm, May to June + July to August), Finnamore Lakes

Burnet Companion, *Euclidia glyphica*, (13-15mm, May to July), Cranberry Bog, Gallen

May - Week 2

All four of these somewhat similar birds are summer visitors from southern Africa, and feed on insects while in flight. Gilbert White notes 'How punctual are these birds in all their proceedings!' Swifts are truly remarkable, relatively large and agile, and of indomitable spirit. They form gregarious, excitable flocks, they mate for life, eating invertebrates, drinking and sleeping on the wing, for up to 15 years. In their lifetime they fly astounding distances, equivalent to the moon and back eight times. Landing only to breed, they return to the same, clean cavity nest each year (unlike the cup-shaped Swallow nest), in buildings or walls. Dark brown-black, their tail forms a shallow 'V', they have short feet (*Apus Apus*, 'without feet') adapted for grasping vertical surfaces (unable to perch on telephone wires), and are found predominantly in towns and villages. We must make every effort to ensure the availability of suitable nesting boxes which can easily be integrated into buildings.

Swallows are a dark, glossy blue, with a deep-red face, creamy-white breast and belly, and a long, forked tail. They are declared the herald of spring when they first arrive in early April.

Swift, *Apus apus*,
(16-18cm long, wingspan 38-40cm,
late April to August)

Swallow, *Hirundo rustica*,
(17-19cm long, wingspan 32-25cm,
March to September)

Step into Nature

Their nests are a familiar sight in sheltered locations under eaves or porches. Bowl-shaped, intricately made of mud predominantly inside old sheds. Unlike the tidier Swift, they do leave guano (bird poop) in the vicinity. They too return to the same nesting site, and also drink on the wing, sipping from streams or ponds.

Another common visitor, the House Martin, is steel-blue-black above, with a white rump, a short, forked tail, and white underside. Smaller than the Swallow but slightly larger than the Sand Martin, they build a bowl-shaped nest under the eaves of houses.

Sand Martins are light-brown with a short, forked tail, white throat and belly, and a thick brown band across breast. They nest in burrows by riverbanks and quarries in rural areas or wetlands and are susceptible to predation by Mink and fox. Our smallest of these *Hirundine* species, they cover large areas when feeding and are often the first of the four species to be encountered in spring.

Sand Martin, *Riparia riparia*, (12cm long, wingspan 26-28cm, April to September)

House Martin, *Delichon urbicum*, (13cm long, wingspan 26-29cm, March to October)

May - Week 3

Week 3

Shady woodlands are home to a petite, umbellifer called Sanicle, with a tall stalk and a basal rosette of leaves. The flower head is made up of outer male and inner female flowers. They are visited by flies and beetles. Historically, Sanicle was sought after to stop bleeding, treat coughs, sore throats and lung complaints.

Sanicle, *Sanicula europaea*, (60cm high, May to July), Mount St Joseph

Next we go to an area of limestone grassland for Mountain Everlasting. Downy, basal rosette leaves are narrow and underside white hairs provide a fetching outline. The flower heads are dioecious, female (12mm) with pink-tipped bracts and male (6mm) with white-tipped, both on a downy stem. Cat's-paw, another common name, comes from the flower shape.

Mountain Everlasting, *Antennaria dioica*, (20cm high, flower 6-12mm, May to August), Clonfinlough

Tormentil sings of summer with solitary flowers of four, bright-yellow petals, slightly notched, brightens meadows, bogs and woodlands. Though similar, Creeping Cinquefoil has five heart-shaped petals on its solitary flowers, found in grassland, hedge banks and waste ground. We must remember to look out for it in September and check for *Fenella nigrita*, its sawfly leafminer.

Tormentil, *Potentilla erecta*, (30cm high, flower 7-15mm, May to September)

Creeping Cinquefoil, *Potentilla reptans*, (20cm high, flower 15-25mm, May to September), *Fenella nigrita* sawfly leafminer

Sometimes old walls or stony ground are enlivened by the fern-like Yellow Corydalis. The long, lipped yellow flowers are carried on branched stems. Great bunches embellish School Lane walls in Crinkill and tiny clumps can be seen along the stone bridge at Tullamore Railway Station.

Yellow Corydalis, *Pseudofumaria lutea*, (30cm high, flower 12-18mm, all year), Crinkill

May - Week 3

Yellow-rattle, *Rhinanthus minor*, (45cm high, flower 13-15mm, May to September), Clonmacnoise

Yellow-rattle, a native annual, is an important wildflower growing in meadows and grassy places. It flowers on leafy spikes with square stems. After pollination (by bees), the enlarged sepals hold the seeds, which rattle in the breeze, giving its common name. Semi-parasitic on grasses and surrounding plants, it is prized for reducing grass growth, allowing an increase in native wild flowers.

By June, an abundance of creamy-white flowers grace riverbanks, bog pools or damp meadows. The almond fragrance of Meadowsweet will fill the air but, for now, we are looking out for the dark-green toothed leaves.

Yellow-rattle, *Rhinanthus minor*, seedheads, Cloghan Lakes

Step into Nature

Some of the leaves have numerous red pimples, made by the gall midge *Dasineura ulmaria*. It vacates the gall on the underside of the leaf, leaving hairy yellow protrusions. There are many fascinating galls on plants and trees caused by insects, mites, fungi and bacteria. Gall mites are tiny insects that cause irregular growths (or galls) which do not generally affect the health of the plant. Some Birch leaves have a furry, cerise pink, felt-like gall, caused by *Acalitus longisetosus*, a *Eriophyde* mite. It develops on the upper leaf between leaf veins. Even tiny Germander Speedwell has its own common gall, *Jaapiella veronicae*, which binds leaves at the shoot tip in a hairy pouch containing several startlingly orange larvae.

Dasineura ulmaria, gall midge on Meadowsweet

Acalitus longisetosus, on Birch

Jaapiella veronicae, gall midge on Germander Speedwell, *Veronica chamaedrys*, garden

May - Week 3

Common Cockchafer, *Melolontha melolontha*, (20-30mm, May to June), with large feathered antenna

Common Pill Beetle, *Byrrhus pilula*, (7-9mm all year), Gorse, Cranberry Bog. For defence, it tucks its legs underneath, resembling animal droppings

Red-headed Cardinal Beetle, *Pyrochroa serraticornis*, (10-14mm, May to July), giving a jaunty wave in the garden

Malachite Beetle, *Malachius bipustulatus*, (5-6mm, May to July), on Cuckooflower, garden

Mid-May is a great time to take a look at some of our beetles. One of the largest and most numerous is the Cockchafer, or May Bug. They are not the best fliers, appearing too heavy for their own wings, as they blunder around gardens and woodlands. Females lay eggs in the ground with a pointed rear section, which hatch into white grubs with a brown head, taking three to four years to pupate.

The attractive moss feeder, a Common Pill Beetle was surveying his kingdom from the tip of a Gorse bud. It has short antennae, a hard, rounded form with a metallic black-purple sheen and scattered dull-gold spots. As with many beetles, when startled, it played dead, very convincingly.

Red-headed Cardinal Beetles add a splash of colour to shaded places in the garden, close to trees. They are red all over, save for smart black legs and long, toothed antennae, similar to the Black-headed Cardinal Beetle, *Pyrochroa coccinea* (but no black head). Eggs are laid in fallen timber and larvae take two years to develop. Another super beetle is the smaller, colourful Malachite Beetle - a beautiful green pollen feeder with prominent red tips to the elytra (wing casing). Its larvae also live in the bark of dead trees or wood, feeding on young insects, pupating in winter.

May - Week 3

Mirid or Capsid bugs are a diverse, cosmopolitan group and the largest family within the Heteroptera order. They may be found on common garden plants, Nettles, Willowherb. *Liocoris tripustulatus* is shiny black or deep chocolate brown with yellow spots. As an adult, it overwinters and is common all year, on flowering plants and Nettles. As with many bugs, it lives on plant sap, sucking on buds, stems, flowers and seeds. *Heterotoma planicornis* is a very distinctive, relative newcomer to Ireland, shiny blue-black with lime green legs and thick antennae. With a varied diet, it feeds on insects, buds and unripe fruit around Nettles, hedgerows and grassland.

An elongated flower bug, with notched knees, *Stenodema laevigata* can be found in semi-damp areas on long grass. If we hunker down, stay still for a moment, we will almost always see a few feeding on the flowering heads, buds or grains of grass. They become active in early April, after hibernation; pair by May with males dying by early June. Eggs are laid by the end of May in grass spikelets, hatching in two weeks. Larvae develop within a month. Oddly, adults change colour from their initial yellow-brown in July to late August. During hibernation, their blood becomes green, giving females a green appearance by the following May.

Step into Nature

Liocoris tripustulatus, (4-5mm, all year), garden

Heterotoma planicornis, (4.5-5.5mm, May to September), garden

Stenodema laevigata, (8-10mm, all year), on long grass, garden, September

Stenodema laevigata, on long grass, garden, May

May - Week 3

Not perhaps the most colourful mirid, *Harpocera thoracica* is the first Oak bug we will see in the spring, in woodlands or gardens. Eggs are laid in young Oak tissue. Larvae hatch after 11 months and live in the scales of partially opened buds, becoming adults by the third week of May. Males are short-lived, (just a week) while females live until mid-June.

A master of disguise, the Birch Catkin Bug blends into catkins from the previous year with its red-brown colouring and transparent forewings. It is found on Birch or Alder catkins in woodlands, parks and gardens.

Known somewhat harshly as being noisy and smelly, it overwinters in the bark of trees and lays eggs on Birch fruits, which take three months to develop.

Harpocera thoracica, (6-7mm, May to June), Birr Demesne

Birch Catkin Bug, *Kleidocerys resedae*, (5mm, all year), on green and dark catkins, Clooneen Bog

Step into Nature

Sometimes, from May to July, we can find delicate Lacehoppers on trees or shrubs. They have long clear forewings supported by an intricate framework of patterned veins. The Spotted Lacehopper, *Tachycixius pilosus*, has a number of dark spots near the wingtips. The larvae feed at the base of grasses around woodlands. *Leucozona lucorum* is a vibrant, well-marked hoverfly with broad cream and black banding to the abdomen, a yellow triangular scutellum and ginger-haired thorax. A bee mimic, it is frequent seen visiting garden flowers and hedgerows on sunny afternoons.

One of our earliest sawflies, *Aglaostigma aucupariae* is widespread in spring. It has distinctive, bright yellow pronotum markings (below head) and a black and red abdomen. Its larvae feed on bedstraw.

Tachycixius pilosus, (5mm, May to July), on Nettle, garden

Leucozona lucorum, (7.75-10mm, May to August), on Cuckooflower, garden

Aglaostigma aucupariae, (7-9mm, March to June), garden

May - Week 3

Mayflies are perhaps best known for their remarkably short (non-feeding) adult life, from just a few hours to two weeks. Nymphs have gills along the length of their abdomen and feed on algae and debris in lakes, rivers and watercourses. As both a nymph in water and an adult, they have long tail filaments (nymphs three, adults two or three). Close to maturity, they develop into a winged sub-imago, or penultimate moult, at the surface of the water. They are the only insects to go through this stage and to moult after gaining wings. They then moult for a final time into the lighter and more elegant adult. Adults have two large front wings and tiny (or absent) hind wings. One example, *Cloeon dipterum,* has a green-brown nymph. The adult has large eyes, two tail filaments and no hind wings. The female may be seen flying over stagnant water or ponds as she lays over 600 living nymphs. They develop over two years, moulting an astounding 24 times. As with dragonfly and damselfly exuvia, the empty sub-imago cases may be found on waterside plants.

Mayfly *Cloeon dipterum*, female imago, Turraun Bog

Cloeon dipterum, sketch of 20mm nymph (top). Male mayfly with turbinate eyes for scanning females above, in mating swarms

Step into Nature

Week 4

It always seems slightly surprising to come across fungi in springtime. Among rotting timbers on damp ground, while seeking emerging dragonflies, (looking from a distance similar to Sundew), we might notice a small, bright red flat fungus. One of the 'disco' group, the Eyelash Fungus has a smooth centre surrounded by an up-turned brown rim, fringed with long, dark bristly hairs.

Eyelash Fungus, *Scutellinia* sp., (10mm, May to November), Turraun Bog

May - Week 4

Fine-leaved Sandwort, *Minuartia hybrida*, (5-15cm high, flower 4mm, May to June), Ridge Road, Birr

Step into Nature

There are some incredibly varied habitats which house a remarkable variety of plants and insects. The Ridge Road near Birr is an esker and the limestone ground provides what is known locally as the 'Mini-Burren'. Fine-leaved Sandwort is an impossibly perfect, low-growing plant. Flowers with five satin-white, widely spaced petals and longer interstitial, pointed sepals.

Aptly named Quaking-grass sways delicately in the breeze while the large white flowers of Burnet Rose brighten the roadside. We are very fortunate to also find Bloody Crane's-bill in this location from late-May into June. Five or seven-lobed leaves are each deeply divided. Large flowers of vibrant pink are produced alone on a long stalk, and weave their way through a striking array of wild plants.

Quaking-grass, *Briza media*, (25-30cm high, May to August)

Burnet Rose, *Rosa spinosissima*, (65cm high, flower 30-50mm, May to June)

Bloody Crane's-bill, *Geranium sanguineum*, (30cm high, flower 30mm, May to June)

May - Week 4

Parent Shieldbug, *Elasmucha grisea*, (7-9mm, all year), the only Shieldbug to mind her eggs, Clooneen Bog

Parent Shieldbug, hatched eggs, Turraun Bog, June, see also July - W4

Step into Nature

Parent Shieldbugs found around Birch, in bogs or the edge of woods, are a patterned mix of browns and red. Bronze Shieldbugs, with slightly pointed shoulders and a yellow mark close to the antennae tip, live in a range of trees, feeding on moths and beetles. The Spiked Shieldbug, with unmistakable pointed, thorn-like shoulders, will be found from July, but we can now look out for nymphs. It favours damp areas with lush vegetation, a formidable carnivore as both larvae and adult feed on moth and sawfly larvae and leaf-eating beetles. All stages also suck plants, if too cool they rest immobile, close to the ground, preferring warm days, sunbathing on the top of plants.

Remarkable Bronze Shieldbug eggs, an iridescent cup with a spiky rim, May, and hatched a month later, garden

Bronze Shieldbug nymph, (July to September), Killaun

Bronze Shieldbug, *Troilus luridus*, (10-12mm, all year), adult, Turraun Bog

Spiked Shieldbug (x2), *Picromerus bidens*, (12-13mm, July to November)

May - Week 4

A spectacularly colourful Ruby-tailed Wasp, *Chrysis* sp., is always a delight to see basking on fence posts on a sunny day. Glistening with a spectrum of metallic colours, the armoured exterior prevents them getting stung by their host, *Ancistrocerus gazella*.

Ruby-tailed Wasp, *Chrysis* sp., (8-10mm, May to September) on garden post, photo for scale

Ancistrocerus gazella, nest building in garden bee hotel, July

Step into Nature

Similar to Nomad bees, the Ruby-tailed Wasp lays her eggs in the host's nest. Then her larvae eat the eggs or larvae of the host and their supply of food. This *Ancistrocerus gazella*, had just started a nest in the garden bee hotel in July.

Two similar beetles often rest on the underside of Birch leaves in the garden. *Malthinus flaveolus* has a black and yellow head, yellow legs and brown elytra with bright yellow tips. *Malthodes marginatus* is predominantly black with two yellow spots.

We came across some unusual lumps on Buckthorn branches. A dark upper casing blended with the branch and covered the softer, white, sap-sucking Scale Insects beneath. These are the apterous (wingless) females. Males (aphid-like) have two large, clear wings. There are many different species depending on the type or absence of outer casing; soft, cushion, nut or woolly.

Malthinus flaveolus, (5-6.5mm, May to August), on Birch, garden

Malthodes marginatus, (4-5. 5mm, June to July), on Birch, garden

Scale Insect, *Coccoidea*, (1-6mm), on Buckthorn, garden

May - Week 4

Fox moth eggs, Turraun, June

Fox moth caterpillar, Finnamore Lake, October

Fox moth, *Macrothylacia rubi*, (22-31mm, May to June), adult, May

Step into Nature

The adult Fox Moth is such a large, luxuriant creature and a delight to meet from May to June. It varies in colour from red-brown to grey-brown with two pale, central cross-lines. If lucky, we may see males flying in the afternoon sun, as they search for females, who, unbeknownst to them, fly only at night.

Glossy, buff-coloured eggs are laid in June, as sticky batches along plant stems. Black caterpillars hatch within a month, have a shiny head, pale orange rings and are covered with long black hairs, flecked with white. The caterpillar continues to grow into the autumn, fading slightly to a dark brown with partial orange bands, until returning to a final velvet black. It is extensively covered in long red-brown hairs and tufts of white hairs down each side.

Feeding until October, it then overwinters until reappearing for any available sunshine at the start of the year. But it no longer feeds at this stage. By March or April, it pupates into a long, brown, cigar-like cocoon of larval hairs combined with silk, fixed low on Heather, grass or moss.

Fox moth caterpillar, Ferbane Bog Walk, July

Fox moth, cocoon, Woodfield Bog, Clara, May

May - Week 4

The bagworm moth, *Psyche casta* - is most unusual. It makes an artful, external casing for protection. Using surrounding grass or reeds, the caterpillar constructs a case which splays out at the end. This self-spun case forms an important long-term home where it pupates. It is also where the blind, limbless, apterous (wingless) female lives out her adult life. A bleak existence for the poor female. Adults do not feed at all, their role is to mate and reproduce.

With less than 60 records across the country, they are not often spotted, despite living in many habitats, particularly light deciduous woodland.

Bagworm, *Pschye casta*, (6-10mm), female inside its case on Whitebeam sapling beside a stream, Kingstown

Pschye casta, (6-7mm, May to July), sketch of adult male

Step into Nature

The small male is dark brown with broad, rounded wings and flies early in the morning.

The Marsh Fritillary flies from May to June in bogs, grassland and eskers. It is a stunning butterfly, beautifully marked with dark orange, cream and contrasting black. The hindwing is more subtle without any black. They may be found basking in early morning sunshine. Females lay batches of yellow eggs on the underside of Devil's-bit Scabious leaves. Within a month, the eggs turn red and hatch. Larvae spin a communal nest in which they can feed and grow in relative safety. By early October, they have moved, together, to fresh plants where they will overwinter beneath vegetation. Towards the end of February, the black spiny larvae are sufficiently warmed up and become active again. By April, they pupate in a dramatic white case with black and yellow markings.

Marsh Fritillary, *Euphydryas aurinia*, Turraun Bog, March

Marsh Fritillary, *Euphydryas aurinia*, (42-47mm, May to June), Turraun, May, and sketch pupa above

June

Week 1 On a damp, uninspiring day, we took the opportunity to visit a new stretch of boardwalk at Clara Bog. There was not much activity, yet in the distance, a glint of blue signalled a vibrant metallic Blue Shieldbug, sparkling with readily forgotten rain. While found throughout Ireland, they are only seen occasionally around the edge of lakes, bogs and marshy grassland.

Blue Shieldbug, *Zicrona caerulea*, (5-7mm, August to June), Clara

Step into Nature

There are two charming tortoise beetles which we can be fortunate enough to come across, low in the undergrowth, on thistles and Burdock. *Cassida rubiginosa* is a handsome oval beetle. It is green with pitted longitudinal patterning, faint golden T-shaped markings and a broad surrounding rim also tinged with gold.

Cassida vibex are found in marshes, bogs or woodlands on thistles, Burdock or Knapweed. They are bright green with a metallic sheen, distinctive dark-brown T-shaped markings and punctured patterns along its upper surface. Eggs are laid on the underside of larval foodplant, which hatch a week later. They continue to feed on the leaves until fully grown around five weeks later and pupate on the leaf surface.

Cassida rubiginosa, (6-8mm, May to August), Clooneen Bog

Cassida vibex, (5.5-8mm, April to September), Clooneen Bog

June - Week 1

We have already been introduced to many different species across a range of families, and as the year continues, we look forward to meeting with further examples. A colourful mirid bug that we come across frequently in the garden or hedgerow is *Grypocoris stysi*. Long, black with a chequered pattern of pale green, yellow and orange; adults and larvae feed on flowers and aphids. Their eggs hatch in May with nymphs reaching adulthood by June or July, surviving for less than eight weeks.

Another Longhorn Beetle, (see Two-banded Longhorn Beetle, April - W4) is the petite Lesser Thorn-tipped Longhorn Beetle. It is dark brown with a light, chevron-shaped marking to the elytra, and dramatic, striped antennae. This fine example was resting on Nettles by the hedgerow.

Grypocoris stysi, (6-8mm, June to August), garden

Lesser Thorn-tipped Longhorn Beetle, *Pogonocherus hispidus*, (4-6mm, May to June), on Nettle, hedgerow

Step into Nature

A familiar hoverfly with its own descriptive common name, is the Marmalade Hoverfly. Depending on the temperature the larvae developed at, the adult *Episyrphus balteatus,* may have a dark (cool) or light (warm) background. A unique colourful hoverfly, it has alternating dark and orange bands to the abdomen. It is widespread and a useful addition to gardens as it feeds on aphids.

Potter Wasps make their nests in a variety of locations. Favouring damp, partially wooded parks or gardens, the solitary wasps build a nest with several compartments. Inside, the female stores moth caterpillars (as food) along with one egg. This typically slender bodied, *Ancistrocerus trifasciatus* was found basking in the garden, with its distinctive three yellow bands. Next, we will take a look at some of the sawflies found nearby.

Marmalade Hoverfly, *Episyphus balteatus* (6-10mm, May to August)

Potter Wasp, *Ancistrocerus trifasciatus*, (7-12mm, June to August), Turraun Bog

June – Week 1, a selection of sawflies

Pristiphora conjugata, gregarious feeders on Aspen, Cranberry Bog, a first record for Ireland, 2023 and recently hatched eggs around leaf tip, with tiny larvae, Finnamore Lakes,

Pristiphora leucopus, early instar, solitary feeders on Lime, Birr Demesne

Pristiphora leucopus, late instar, on Lime at Birr Demesne, a first for Ireland, 2022

Step into Nature

Formerly known as *Abia sericea*, *Abia nitens* (10mm, May to August) adult on Birch, Finnamore Lakes, May and its larva found (eventually, after a considerable search), munching Devil's-bit Scabious, August

Eriocampa ovata, adult (5-7mm, May to August) on Grey Alder, garden, May and larva (designed to look like bird poop) underside Grey Alder, garden, October

Nematinus steini, (5-8.5mm, June to August) female ovipositing on Grey Alder, garden

Nematinus steini, larva on Grey Alder, garden, July

June - Week 1

Common Figwort is an inconspicuous tall plant of damp, shady areas. Loose clusters of small green and purple flowers rise from square stems. Visited predominantly by wasps, when flowers ripen, their fruits look like tiny figs. Resting on the flowers, we may find a small Figwort Weevil with a seemingly oversized abdomen. It has raised white-spotted ribs, a small white prothorax (bears first set of legs) and a long, curved snout. What appears to be a hole in its back is actually a large black spot. As with many beetles, when startled it tips off the edge of the plant, disappearing to the ground below.

There are many similar yellow flowers (Hawkweed and Hawkbit), but Goat's-beard really stands out. Its leaves are long and narrow, elongated buds encased in eight elegant blue-green bracts, but it is the large seed head that is truly stunning.

Common Figwort, *Scrophularia nodosa*, (1m high, June to September), and Figwort Weevil, *Cionus scrophulariae*, (4-5mm, June to September)

Goat's-beard, *Tragopogon pratensis*, (60cm high, flower 20-30mm, May to August), bud and seed head, Ridge Road

Step into Nature

A much larger globe than the more common Dandelion, each of the intricate pappus appear almost concave, giving the appearance of a delicate, modern light fitting. Known as Jack-go-to-bed-at-Noon as on either a dull or sunny day, the flowers close at midday.

A flower which combines stunning beauty with a bizarre design, colour and texture, is the fantastical Bee Orchid. It favours rough, gravelly ground such as eskers or quarries and is a glorious sight to behold. A sight to make your heart dance.

Bee Orchid, *Ophrys apifera*, (40cm high, flower c.30mm, June to July), Ridge Road, Birr

June - Week 1

Poplar Hawk-moth, *Laothoe populi*, (36-44mm, May to July), adults in garden, June, and larva (June to October), feeding on Willow at Turraun, August

Ghost Moth, *Hepialus humuli*, (21-35mm, June to August), female yellow-orange, male white, garden

Step into Nature

The large, satin-textured Poplar Hawk-moth overwinters as a pupa in the ground close to Willow and Apple. Its equally impressive bright green caterpillar has diagonal yellow stripes and a quirky yellow horn.

There can be subtle or great differences between male and female moths. The male Ghost Moth merits its name due to white wings, and its ghostly dance at dusk to attract females. Females are soft gold with orange markings. Some moths, like their caterpillars, are designed to go unnoticed and a very endearing example is the Chinese Character. As it rests by day on vegetation, it looks so like bird droppings that it can easily be overlooked. We have also found their small bud-like larva feeding on Blackthorn in the garden.

Chinese Character larva on Blackthorn, garden, September

Chinese Character, *Cilix glaucata*, (10-13mm, April to June and July to September), at rest in the garden

June - Week 1

We do not always need a trap to see adult moths as there are some superb varieties to be found flying during the day. Common in gardens and woodlands, Angle Shades may be found all year. Fresh examples have a beautiful range of olive-green to pink and brown and rest with folded wings, like leaves. They overwinter as a green or brown larva, also found all year. Another garden moth, similar to that found on Dame's Violet (March - W3), is the Diamond-back Moth. Located close to its larval foodplant (cabbage family), it is a delicate yet noble little garden visitor.

The Nettle-tap has an oscillating flight as it flits along the hedgerow, seemingly never intending to land. It is a petite yet sumptuous moth of rich chocolate browns. Adults have at least two broods each year. The caterpillar forms a web on the upper side of Nettle leaves and it draws the leaf closed for security.

Angle Shades, *Phlogophora meticulosa*, (21-25mm, all year), garden

Diamond-back, *Plutella xylostella*, (6-8.5mm, all year), Killaun Bog

Nettle-tap, *Anthophila fabriciana*, (5-7mm, April to November), garden

Step into Nature

From May to July, if we go to our local woodland or bog, we will be sure to see the Marbled White Spot. If disturbed, while we walk, it flies off ahead coming to rest on vegetation. Pupae overwinter below ground and the pale brown or green larvae feed on grasses.

The unmistakable and beautifully marked Silver Hook also flies when disturbed, but not too far. With cautious movement, we may observe it settling nearby and it is well worth taking the time to catch up with this unique moth. It flies in bogs, marshes and fens by day but particularly from dusk.

This is the perfect time to observe the Six-spot Burnet across areas of grassland. Often fresh out of yellow cocoons still fixed to grass stems, they flutter short distances in search of a mate. Great numbers appear as a blur of vivid black and red. The black and yellow caterpillars feed on Bird's-foot-trefoil.

Marbled White Spot, *Protodeltote pygarga*, (11-12mm, May to July), Turraun Bog
Silver Hook, *Deltote uncula*, (11-12mm, May to July), Burren National Park
Six-spot Burnets, *Zygaena filipendulae*, (17mm, June to August), and cocoon, Birr Demesne

June - Week 2

A dainty Small Heath,
Coenonympha pamphilus,
(33-37mm, May to August)

180

Step into Nature

Week 2

With balletic grace in sunny weather, the Small Heath butterfly flies swiftly over grassland, meadows and eskers. It lives as an adult for just one week. Eggs are laid on dead blades of grass and hatch a week later. Yellow-green larvae feed on grasses in sunny weather, rest at the base by night and can hibernate during the winter months. Squat yellow pupae develop into a vibrant green between April and August. While trying to get a (non-blurred) photograph of the speedy butterfly, we spot some strange galls on Blackthorn. The more conspicuous 'pocket plums', *Taphrina pruni*, form where an ascomycete fungus (spores formed in a sac) attacks the sloes (fruits). This destroys the stone and seed. Spores develop on the outside, infect the stems and start again the following year. Small pink pimples on Blackthorn leaves signal the gall mite *Eriophyes similis*, which is found from spring to autumn.

Taphrina pruni, (May to September), ascomycete fungus on Blackthorn

Eriophyes similis, (May to September, gall mite on Blackthorn, garden

June - Week 2

Both of these hoverflies are unusual, as even in flight, they are fortunately unmistakable. This particularly smart *Chrysotoxum bicinctum* rested on Birch at Turraun Bog, along a woodland walkway. They are also found in open grassland. A wasp mimic, it has distinguishing yellow bands on a relatively squat, broad abdomen with dusky chocolate markings to the wings, and notably long antennae (for a hoverfly).

A crisply marked *Chrysotoxum bicinctum*, (7-10mm, June to September), on Birch at Turraun Bog

Step into Nature

Another very attractive hoverfly found in the garden is *Scaeva pyrastri*. This is a migrant species that some years is totally absent, perhaps due to weather conditions. The black abdomen has six white comma-shaped markings (these are yellow in other related species). The slug-like larvae of most hoverflies feed on garden aphids.

Scaeva pyrastri, (9-12mm, June to August), on Sweet William, garden, and hoverfly larva on White Campion

June - Week 2

Green-winged Orchid, *Anacamptis morio*, (30cm high, flower 8-12mm, April to June), Callows, Crank Road

Early Marsh-orchid, *Dactylorhiza incarnata agg.*, (30cm high, flower 8-12mm, May to July)

The scarce Green-winged (or Green-veined) Orchid may be found mainly in the Midlands within short, esker grassland. Leaves form a basal rosette and sheath the stem. But its most distinguishing feature is the green veins on the lateral sepals. Flowers are typically purple to magenta, while occasional pale pink or white varieties really highlight the green veins. They are pollinated by bees. Visually similar to Early Purple Orchids, their timeframe also overlaps.

Early Marsh-orchids are found by lakes, bogs and marshes. They are generally purple to mauve with unspotted, folded leaves. Within the Shannon Callows, they can grow close to Green-winged Orchids.

Step into Nature

Pyramidal Orchids always appear as perfectly petite, pink pyramids (cones). They have lanceolate leaves which sheath the stem and deep-pink flowers. Found in grassy, limestone locations, they are relatively common. They are pollinated by moths with a skilful mechanism, that Darwin felt was better perfected than in any other plant.

Pyramidal Orchid, *Anacamptis pyramidalis*, (50cm high, flower 3-6mm, June to August), Lough Boora

The Lesser Butterfly-orchid is yellow-green to white, and grows in bogs and heaths. As with Honeysuckle, it releases its scent in the evening because it is pollinated predominantly by moths. A similar Greater Butterfly-orchid favours woods and grassy places.

Lesser Butterfly-orchid, *Platanthera bifolia*, (30cm high, flower 10-15mm, May to July), Clara Bog

June - Week 2

We have come across a few sawflies already and know that they can vary in size, but some, like *Cimbex femoratus* (May - W2), really are huge. Returning from a walk one day, we noticed a very large bee or wasp type insect on the doorstep. It was a remarkable *Cimbex connatus* whose larvae feed on Alder. What an exceptional encounter! A yellow and brown striped sawfly, it is similar to *Cimbex luteus* which is not yet recorded in Ireland. However *Cimbex connatus* has a darker orange thorax and a shinier, hairless scutellum.

The vast, magnificent *Cimbex connatus*, (20-28mm, May to June), temporarily settled on the doorstep, June

Step into Nature

Some weeks later (after lots of anticipative searching), a single delightful larva was found in its customary resting position, on the underside of a Grey Alder leaf, a truly proud moment that the lifecycle had been realised in our own garden. Larvae are yellow-green with dark blue and raised white spots and a dark dorsal stripe.

The species is spreading again across England and Ireland since 1997, after an earlier absence.

Cimbex connatus, (larva up to 50mm, July to September), a single larva on Grey Alder, garden, July

June – Week 2

Bracken is a familiar site in hedgerows and bogs. This is the perfect time to check unfurling fronds for the distinctive, long sawfly *Strongylogaster multifasciata*. Females oviposit into the Bracken and two weeks later, her green larvae hatch, with a black spot on each temple. Other current sawflies include the pale-legged *Heterarthrus nemoratus*, the social *Hemichroa crocea* larvae in the garden (still searching for the orange-red adult) and an elegant *Allantus* sp., whose green larvae feed on Brambles and roses.

Strongylogaster multifasciata, (9-11mm, May to September), ovipositing on Bracken

Strongylogaster multifasciata, adult 31 May and larva 18 June, same Bracken, Killaun Bog

Step into Nature

Heterarthrus nemoratus, (4-6mm, May to July) female (males unknown), on Birch, Pallas Lake

Possible *Heterarthrus nemoratus* leafmine, same Birch two weeks later

Hemichroa crocea, (adults 5-8mm, May to September), larva on Birch, garden

Allantus sp., female (7-10mm, May to August), Clooneen Bog

Hemichroa crocea, recently moulted larvae with exoskeleton below, on Birch, garden, also found on Alder

June - Week 2

Limnephilus lunatus, (10-15mm, April to November), moth trap, garden

Limnephilus auricula, (8-12mm, April to October), moth trap, garden

Glyphotaelius pellucidus, (12-17mm, May to October), trap, garden

Mystacides azurea, (6-9mm, May to October), by day, Cranberry Bog

Earlier in the year, we discovered the incredible cases made by the larva of the caddisfly, February - W3. Now it is time to meet some of the vast range of adults found at a variety of water-bodies. One of the easier caddisflies to identify (and one of the most common) is *Limnephilus lunatus*. It is part of the largest caddisfly genus, with 29 species in Britain and Ireland. So named because of the moon-shaped or crescent marking at the apex (tip) of the forewing. *Limnephilus auricula* is darker brown with pale markings around the centre of the forewing.

In contrast, *Glyphotaelius pellucidus* is the only species in its group. The forewing notch is distinctive as is the brown and cream marbling.

Opposite page, Caddisfly eggs on Buckthorn, October, Burren National Park

Step into Nature

Eggs are laid in a jelly mass on leaves above water. The jelly swells and after rain, or heavy dew tiny developing larvae drop down into the water below. Larvae then construct an intricate case of fallen leaves.

The sleek *Mystacides azurea* is the angler's Black Silverhorn, with strange red eyes, 'elbowed' maxillary palps (appendages near the mouth), white legs and long white antennae.

Other longhorn sedges can be a spectacular sight, resting on vegetation by ponds, lakes or canals. Often, they are large, pale with phenomenal antennae such as this one, *Oecetis ochracea*, found at Pallas Lake.

An superb longhorn caddisfly, *Oecetis ochracea*, (6-13mm, May to September), by day at Pallas Lake

June – Week 2

Our largest caddisfly, *Phryganea* sp. has a vast forewing length of up to 28mm, but it requires more than a photograph to identify. They have mottled brown forewings and live close to still waters in ponds, lakes, canals and slow rivers. The female is larger than the male and she usually has a dark black band on her forewing. She crawls under the water to lay a ring-shaped jelly mass of eggs on plant stems. Larvae, up to 40mm long, construct a spiral case of uniformly sized leaves or small sticks.

Stenophylax permistus live close to pools or streams and disperse widely from their breeding sites. The attractively–marked *Hagenella clathrata* is a rare find by tiny pools on raised bogs and heaths as there are just a few Irish records. All three were found resting by day.

Phryganea sp., (up to 28mm, May to September), Turraun Bog

Stenophylax permistus, (18-24mm, March to October), garden

Hagenella clathrata, (11-15mm, June to July), Turraun Bog

Step into Nature

Alderflies (of the small order Megaloptera), are strong-looking dark insects with a broad head and two pairs of black-brown, thickly-veined wings. These are held closed 'tent-like', unlike Stoneflies (November - W4) who rest with wings held flat. Alderflies are awkward, apathetic fliers preferring to rest on vegetation by lakes, ponds, rivers and streams.

Eggs masses are laid with angular precision on waterside vegetation. Larvae hatch and drop into the water where they feed on crustaceans, worms and insect larvae for almost two years. Short-lived adults emerge in early summer for just 2-3 days. Their sole focus is to mate and lay eggs. There are only two species in Ireland; *Sialis lutaria* and *Sialis nigripes*.

Sialis sp., (body 10mm, wingspan 22-34mm, May to June), May, laying eggs at Finnamore Lakes

June - Week 3

Week 3

Some very interesting plants are now coming into flower across bogs or damp meadows. Marsh Cinquefoil of the Potentilla family, is tall and very striking, with its deep-wine flowers of five sepals and elegant buds. Every year, by the boardwalk at Killaun Bog, we find the noble, and aptly named Royal Fern. Large, broad yellow-green fronds (fern leaves) rise up to an impressive 180cm. What appear like rusty flowers are fertile fronds full of merging sori. Tall, elegant Meadow Thistle has beautiful, soft, pink flowers, held aloft over unbranched stems. For those fortunate to live close to a bog, there can be few more breathtaking sights than the tapering yellow spikes of Bog Asphodel, stretching as far as the eye can see. By Autumn, the flowers have turned orange, continuing to provide a most picturesque effect.

Marsh Cinquefoil, *Comarum palustre*, (35cm high, flower 20-25mm, May to July)

Royal Fern, *Osmunda regalis*, (up to 180cm high, June to August), all at Killaun Bog

Meadow Thistle, *Cirsium dissectum*, (75cm high, flower 15-25mm, June to July)

Step into Nature

195 Bog Asphodel, *Narthecium ossifragum*, (30cm high, flower c.13mm, June to August), Killaun Bog

June - Week 3

There are other damp areas to check for plants, including lake margins and some canals. We may notice the fragile flowers of Lesser Water-plantain. They are three-petalled, pale-pink with a yellow centre and grow out of the mud by the edge of lakes or canals. Common Valerian also favours damp ground by rivers, canals and wet woodlands. We have found it at Turraun Bog, Cloghan Lakes, Clonmacnoise and in Birr Demesne. Not particularly common, as its name might suggest, it has clusters of pink flowers. The more vigorous, darker variety, Red Valerian (which also comes in white) is found by roadsides and on waste ground.

Skullcap, a small, somewhat rare plant grows by the edge of Pallas Lake. Paired blue-purple flowers have a soft, almost velvet appearance to their galeate (helmet-shaped) flowers. Bog Pimpernel is low-growing and favours wet, muddy areas. Stunning bell-shaped, pale-pink flowers are finely veined with red. One of the more well-known plants of damp areas is the aptly named, Ragged Robin. Of the stitchwort family, it brightens marshy ground with its airy, tousled rose-pink flowers.

Ragged Robin, *Lychnis flos-cuculi*, (65cm high, flower 30-40mm, May to August)

Step into Nature

Lesser Water-plantain, *Baldellia ranunculoides*, (18cm high, flowers c.14mm, June to September)

Common Valerian, *Valeriana officinalis*, (1.5m high, flower 5mm, June to August)

Skullcap, *Scutellaria galericulata*, (18cm high, flower c.9mm long, June to August), Pallas Lake

Bog Pimpernel, *Anagallis tenella*, (flower 6-10mm, June to August), Pallas Lake

June - Week 3

With but a modest palette of browns and grey, the Clouded Drab moth is extremely variable. It is the pale green caterpillar that we can expect to see at this time of year. Speckled with white, yellow or bluish-green, it has a dorsal stripe (along its back), and side stripes of yellow, white and sometimes black. So even the caterpillar is variable.

The Magpie moth is such a striking moth which may be found by day, resting on low foliage. Feeding on a range of plants, we found the caterpillar on Spindle. It overwinters as a small larva, growing in size from the previous August until the following June. Initially much darker, it develops into a whiter base with black and orange markings. One of the few to resemble its adult colours.

Top: Clouded Drab, *Orthosia incerta*, (16-20mm, March to May), adult and caterpillar, on garden Abutilon

Magpie, *Abraxas grossulariata*, (18-25mm, June-August), adult, July and caterpillar, March to May garden

Step into Nature

It is always lovely to see the dark fluttering Ringlet butterfly in grassland or woodland. It feeds on Brambles, thistles and Pyramidal Orchids, while larvae feed on grasses. The Large White is recognised by its wingtips - with black extending equally in both directions. On the similar Small White, the dark wing marking extends further up the wing than down. Found in most habitats, Large White larvae are vigorous, gregarious feeders on cabbage plants and Nasturtium. Adults feed on garden flowers including Buddleia, Knapweed and thistles.

The Meadow Brown has dark upper wings with a splash of orange around the eyespot. A grassland butterfly favouring wildflowers, females lay eggs singly on grass, which hatch two weeks later. Hairy green larvae, with six instars, feed on a range of grasses. We have been fortunate lately to have them in good numbers in the garden.

Ringlet, *Aphantopus hyperantus*, (42-52mm, mid-June to August), Turraun

Large white, *Pieris brassicae*, (63-76mm, April to September), on garden Scabious

Meadow Brown, *Maniola jurtina*, (52-56mm, June to September), garden

June - Week 3

Painted Lady, *Vanessa cardui*, (58-74mm, May to September), garden

A few years ago, we noticed a new arrival in the garden seeking nectar-rich flowers. Painted Lady butterflies make their way to Ireland from Morocco each year. As temperatures rise, the powerful, fast-flying butterflies leave the deserts of North Africa, crossing the Mediterranean into Spain. Successive generations fly further northwards across Europe to Ireland as annual migrants, some years they are plentiful others, very scarce. They breed during our cooler summers with their caterpillars feeding on thistles.

Step into Nature

There are five species of Grasshopper (Orthoptera) native to Ireland, two Groundhoppers and five Bush Crickets. Grasshoppers are powerful insects with strong back legs built for jumping. Like many insects they go through several nymph stages before reaching maturity. While grasshoppers are plant-eaters, crickets are mostly predators. The Mottled Grasshopper is a grassland and sandy ground species. It has the dramatic appearance of something loosely pieced together, not quite finished. Colours range from green, brown, purple to black and it has distinctive clubbed antennae.

Irish grasshoppers include: Field, Common Green, Large Marsh, Lesser Marsh and Mottled Grasshopper

Mottled Grasshopper, *Myrmeleotettix maculatus*, (12-19mm, June to mid-September), Clooneen Bog

June - Week 3

Mirid bugs of different shapes and sizes are now appearing, including the distinctive *Campyloneura virgula*. It is a long predatory bug found on Oak, Hazel and Hawthorn. Also long and predatory, *Dicyphus errans* is widespread on a variety of herbaceous plants and an annual visitor to the garden. The substantial *Leptopterna dolabrata* is strikingly marked and may be found in damp areas feeding on a variety of grasses.

Capsus ater is a smart, common grassland bug; dark, brown to black and oval shaped with a brown or black head. It feeds low down on the stems of grasses rather than on the flower spike.

Another slightly oval-shaped bug, *Heterocordylus tibialis*, can be found with some

Campyloneura virgula, (4-5mm, June to October), on Oak, garden

Dicyphus errans, (4.5-5mm, June to October), garden

Leptopterna dolabrata, (8-8.5mm, June to September), garden

Step into Nature

Capsus ater, (5-6mm, June to September), Pallas Lake

Heterocordylus tibialis, (4.5-5.5mm, May to July), resting on Gorse, Cranberry Bog, Gallen

Ditropis pteridis, (3-4mm, May to August), a striped female, (males all-black) on Bracken, Killaun Bog

confidence, when standing still beside its only foodplant, Broom. By focusing on the stems, we can see several, scurrying up and down. But they are very shy and zip away to another part of the plant as soon as we move. Also dark, it is covered in fine hairs that appear almost golden in sunlight.

We should check through Bracken in verges or along woodland walks, close to tree cover. *Ditropis pteridis* is such a tiny and singular bug. With short undersized wings, extending only down the top third of their striped abdomen, females resemble the curled fronds of Bracken on which they reside. Males are smaller (yet conversely have larger wings) with all-black abdomens.

June - Week 3

The legally protected, native Pine Marten is one of our rarest mammals, predominantly of woodlands. The same size as the domestic cat, it is a glossy dark-brown with a creamy-yellow to orange throat and breast with large, bright, black eyes, broad, open, rounded ears, sharp claws and a long bushy tail. They have a mixed (appetising) diet from rats, mice, rabbits and poultry to strawberries, raspberries and bilberries and occasionally honey. Females make nests in tree hollows or old crow's nests, and have two litters a year, of two to seven kits.

The much smaller and significantly lighter Irish Stoat, also protected, is a subspecies of those across Europe. Its upperparts are red-brown while its underparts are creamy coloured throughout from chin to toes. It hunts rabbits, poultry and fish, and is largely, but not exclusively nocturnal. Females give birth to four or five young, April to May, in a tree or bank hollow.

Pine Marten, *Martes martes*, (c.60cm), cat-sized with a creamy cravat to just below the chin

Irish Stoat, *Mustela erminea subsp. Hibernica*, (20-35cm), looking inquisitive in the Burren, with all-white underparts

Step into Nature

Otters live along our coast, rivers and lake shores and we have been fortunate to come across them at Birr Demesne. But an encounter at the Cranberry Bog, Gallen, was even more memorable. On a warm, still late-afternoon, the silence was broken with the slow, languorous sound of something substantial swimming in the adjoining bog pond. A beautiful sleek Otter glided through the water, dipping gracefully up and down as it swam to and from the shore.

Otters have a broad head with a short face, small bright eyes and rounded, hairy ears. Short legs are very powerful and their webbed feet have five toes with short pointed claws. Some have been seen on river banks, curled up with their tail wrapped around them, like a sleeping dog. They hunt fish, frogs and water foul from sunset through the night. A holt (nest) is built of rushes and may be lined with reed flower heads, where the female gives birth to two or three blind young. They are fed by both parents until old enough to be brought out at night to hunt. The male then leaves while the female stays with them until the following year.

Otter, *Lutra lutra*, (up to 1m long)

June – Week 4

Week 4

We have met with some magnificent moths during the month of June, both at Killaun Bog and in the garden, some evenings over 250 moths and 50 different species. A small sample of these incredible creatures follows.

Swallow-tailed Moth, *Ourapteryx sambucaria*, (22-30mm, June to August), unmistakable, pale lemon-yellow

Light Emerald, *Campaea margaritaria*, (18-26mm, May to October), beautiful, pale white-green with a dark-edged white cross-line

A master of disguise, the Peppered Moth, *Biston betularia*, (22-28mm, May to August), on Birch, Killaun Bog

Burnished Brass, *Diachrysia chrysitis*, (16-19mm, June to October), brown with stunning brassy yellow green

Step into Nature

Lilac Beauty, *Apeira syringaria*, (19-22mm, June to July), wonderful raised forewing resting posture and exquisite mix of orange, brown and lilac markings

Large Emerald, *Geometra papilionaria*, (24-29mm, June to August)

A wonderful, quirky, big green moth often found in numbers resting on Heather by the trap

June - Week 4

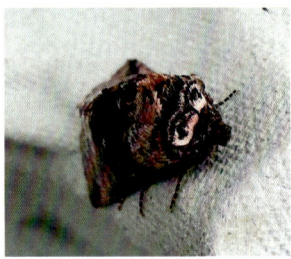

Grey Arches, *Polia nebulosa*, (21-26mm, June to July), well-camouflaged with softly understated tones of cool grey, Killaun Bog

Spectacle, *Abrostola tripartita*, (15-17mm, April to September), quirky with white spectacles topped by a furry mohawk

Dot Moth, *Melanchra persicariae*, (16-21mm, May to June), glossy, black with white kidney mark, Killaun Bog. Caterpillar (August to October), green or brown chevron pattern, raised frontal section and hump towards rear, feeds on herbaceous or woody plants, Alder, garden

Muslin Footman, *Nudaria mundana*, (10-12mm, June to August), a small, delicate, transparent macro-moth. Caterpillars (August to June), are gregarious feeders of lichen on stone walls, (particularly early damp mornings) where they also bask in the sun, garden and the Burren

Step into Nature

Elephant Hawk-moth, *Deilephila elpenor*, (28-33mm, May to August), outrageously pink with olive-green. Caterpillar (June to September), feed on Rosebay Willowherb by night, rests on stems on fine afternoons. Green, turning brown with large 'eyes' to deter predators

Eyed Hawk-moth, *Smerinthus ocellata*, (36-44mm, May to July), startling hindwing eyespots, exposed when disturbed to deter birds. Large green caterpillar (June to October), with white or yellow diagonal stripes and a blue horn (tail) Killaun and Cranberry Bog

Ruby Tiger, *Phragmatobia fuliginosa*, (14-19mm, April to September), richly coloured common species. Caterpillar (July to May), unusually is similar in colouring to the adult with long tufts of swept back hair, feeds on herbaceous and woody plants, Killaun Bog

June - Week 4

Xylota segnis, (7-10mm, May to August), garden

Volucella bombylans, (8-14mm, May to July), garden

Eristalis horticola, (8-11mm, April to October), Birr Demesne

Xylota segnis, one of the elongate hoverflies, distinguished by its partly yellow legs. It may be seen collecting pollen and honeydew as it sweeps backwards and forwards over leaves.

The wonderful bumblebee mimic *Volucella bombylans* is polymorphic (different adult forms). Depending on which bee it is attempting to imitate, it can have a black form with a red tail (Red-tailed Bumble Bee), black and yellow (White-tailed Bumblebee) and a rarer all-buff form on the Aran Islands (Common Carder Bee). This fascinating ability is to deter predators into thinking they could be stung.

The crisply marked *Eristalis horticola* has a dark mark across the centre of its wings. It is found close to bushes or trees.

Step into Nature

Some insects have a different approach to creating a suitable environment for their developing offspring. This 14-spot ladybird has been parasitised by the wasp *Dinocampus coccinellae*. The wasp's egg, laid in the underbelly of the ladybird, takes up to a month to mature. It then emerges (after paralysing the host) and pupates in a cocoon attached to the ladybird's leg. Just over a week later, the adult wasp emerges and, allegedly, 25% of the ladybirds revive.

This photo shows the cocoon underneath which has popped open allowing the adult wasp to emerge.

14-spot Ladybird, *Propylea quattuordecimpunctata*, (3.5-4.5mm), with cocoon of wasp *Dinocampus coccinellae*, garden

Sketch of parasitic wasp *Dinocampus coccinellae*, (3-4mm, spring and summer)

June - Week 4

Banded Demoiselle, male,
Calopteryx splendens, (length 45mm,
end of May to August), Birr Demesne

Step into Nature

There are two Irish species of demoiselles, which are the largest damselflies in Ireland. The Banded Demoiselle is on the wing first. Females are gleaming metallic green which blends into a bronze tail and green tinted wings with a pale white spot, (the Beautiful Demoiselle, also metallic green, has noticeably brown wings). Stunning males are metallic blue-green and have a dark 'fingerprint' on their wings. They are such an attractive species and can be seen fluttering along the banks of slow-moving rivers and ditches, gracefully rising up to rest on riverside vegetation. We should check south-facing banks in the roots of submerged vegetation for stick-like larvae which overwinter in the muddy base for two years. Adults emerge from May to August.

Beautiful Demoiselle, *Calopteryx virgo*, (45-49mm, May to August), Killyon

June - Week 4

Red filaments flicked from the tail to ward off predators

The caterpillar at rest, showing the unbelievable camouflage provided against the veined Aspen leaf

Step into Nature

Remember the beautiful Puss Moth, *Cerura vinula*, we saw in the garden (March - W4)? Well, this is the astounding, late-instar caterpillar. Two were spotted munching on Aspen leaves at the Cranberry Bog, Gallen. It was mesmerising to watch their purposeful progression, nibbling first the leaf, then the stem, as they reversed back.

A week later, the substantial caterpillars are even larger and heavier as they make their way up through the stems

Front feet for gripping the narrow stem as they nibble the stripped leaf stalk, rear feet shuffle up and down the stem

June - Week 4

See also Two-banded Longhorn Beetle, April - W4, and Lesser Thorn-tipped Longhorn Beetle, June - W1

Four-banded Longhorn Beetle, *Leptura quadrifasciata*, (body length 15-27mm, June to July), Turraun Bog

Step into Nature

How spectacular is this dramatic beetle? With quiet magnificence, the Four-banded Longhorn Beetle, with large antennae, fills the listing leaf on which it rests. It is vibrantly marked with four rectangular golden bands to each side of the elytra, set against a black background. What a treat to spot this one at Turraun Bog, who stayed perfectly still, allowing a photograph. Females lay eggs in crumbling, dead deciduous wood, where the pale, thin-skinned larvae will live. Larvae gnaw passageways through timber with their sharp mandibles, later pupating in the tree. Adults then make an oval exit hole from the timber. They are active during the day, particularly if sunny when they may be seen in flowers seeking pollen.

The smaller Garden Chafer is an attractive and distinctively hairy beetle. The red-brown elytra and metallic green or blue head and prothorax shimmer in the sunshine, as they fly past with a somewhat noisy, flight landing in a clumsy manner. Common on bushes and deciduous trees in hedgerows, grassland and gardens, it nibbles holes in the surface of leaves, while larva live in the ground feeding on roots.

Garden Chafer, *Phyllopertha horticola*, (7-12mm, May to June), Ridge Road and garden

July

Week 1 It is a conflicting sensation watching the strong, sweeping flight of Brown Hawkers. Enthralling to gaze as they swoop and glide high overhead above lakes and fens. But frustrating because they are quite so fast and high. Occasionally, they land close at hand, usually busy munching prey. They have brown-tinted wings and a brown abdomen with two striking yellow-white stripes to the side of the thorax.

In contrasting scale, Emerald Damselflies are the embodiment of delicate beauty. They are a wonderful metallic green with clear wings, held partly open at rest. Females breed around acid ponds and small lakes, found at Turraun, Pallas Lake, Cranberry Bog and Lough Boora.

Emerald Damselfly, *Lestes sponsa*, (38mm, June to September)

Step into Nature

Brown Hawker, *Aeshna grandis*,
(73mm, mid-June to September),
Turraun and Cranberry Bog, Gallen

July - Week 1

Ruddy Darter, *Sympetrum sanguineum*, (34mm, June to September), Turraun, with waisted abdomen and black legs (although golden-yellow females do not have a waisted abdomen). Found at fens, bogs, lakes and turloughs

The similar yet larger, Common Darter, *Sympetrum striolatum*, (40mm, June to October), has a straight abdomen, (no waist). Legs, while also dark, have a pale stripe. Females are duller than Ruddy Darter, more of a yellow-brown

Larger again, Black-tailed Skimmer, *Orthetrum cancellatum*, (44-49mm, May to August), Clooneen Bog. Males are a distinctive powder blue with a dark tip to their abdomen

Immature male, Black-tailed Skimmer. They breed at shallow, limestone lakes, fly low to hunt and bask on stony ground, Turraun Bog

Step into Nature

As children, we are used to seeing small, pendulous collections of frothy bubbles hanging from herbaceous plants. Commonly called cuckoo-spit, as they appear in May, at the same time as the Cuckoo. They are made by the nymph of a froghopper (or treehopper) - small spittlebugs with short antennae, tent-like ineffectual wings and strong legs for jumping. Adult females oviposit overwintering eggs on the selected plant. The spit is a form of fluid excrement produced by the developing nymph, mixed with wax and puffed up until it turns to foam (May to June). At this stage, the nymph is soft and vulnerable to both the elements and predators and the foam keeps it from drying out, while obscuring the nymph.

Common Froghoppers vary in colour from yellow to grey to black. Their nymphs are light yellow-green and rest, head down within their foamy shelter. When sufficiently grown, they jump up the stem where they will feed on sap, sucked up with their proboscis.

Common Froghopper, *Philaenus spumarius*, (5-6mm, June to October), garden

July - Week 1

A bank or meadow of tall nodding Field Scabious is such a beautiful sight. Each composite flower or domed head contains about 50 individual flowers, which develop from the outside in (smaller in the centre). They were used to treat a vast range of ailments from scabby conditions, wounds, dandruff and even leprosy. Often unnoticed in bogs or the woodland floor, Cow-wheat seems scarcely able to hold itself upright. Like Lousewort, it is semi-parasitic on grasses. It has unusual tubular flowers, nectar filled and pollinated by bumblebees, while its seeds are dispersed by ants. Of concern to millers, seeds were said to turn flour black and narcotic. Sometimes we encounter perfect beauty in miniature form - Common Centaury, with five, rose-pink pointed, satin petals and central orange anthers. While they only flower for five days, they are discerning, refusing to open in bad weather and when fine, only for a few hours in the morning.

Field Scabious, *Knautia arvensi*, (70cm high, flower head 30-34mm), Red-tailed Bumblebee
Cow-wheat, *Melampyrum pratense* (35cm high, flower 15-20mm). Common Centaury,
Centaurium erythraea, (50cm high, flower c.10mm) all flower June to September

Step into Nature

Smaller and even more delicious with an eggy centre, Eyebright, *Euphrasia nemorosa*, (25cm high, flower c.8mm, May to September). Similar to Cow-wheat, drawing nutrients from grasses, they differ in that they use them to manufacture their own food, making them hemi-parasites.

July - Week 1

Garden Bumblebee, *Bombus hortorum*, (13-18mm, April to May)

Ectemnius lapidarius, (7-12mm, June to September)

The Garden Bumblebee, (March - W2) with its long, 'horse-like' face, can also be recognised by the bright yellow collar and matching central band. It is found in most habitats and gardens.

The square-headed wasps, *Ectemnius lapidarius*, live close to dead wood, in woodland, scrub and wetlands. Females burrow into tree stumps or fence posts, placing a single egg in cells constructed along either side of a central burrow. A paralysed hoverfly is left in the cell and is consumed by the hatched wasp larvae. The larvae then pupate and emerge the following year. Next, let us look at two elegant and similar hoverflies, found on umbellifers such as Hogweed.

They are both woodland species and of the same family as *Leucozona lucorum,* which we found in May - W3, yet quite different. *Leucozona laternaria* have dark front legs and a dark scutellum which leads into a deep white collar on the abdomen. While in the similar *Leucozona glaucia,* the stripes have a pale blue tinge and are separated with yellow between the scutellum and abdomen.

Volucella pellucens is a recognisable hoverfly, with a shiny black scutellum and an abdomen of one broad black and white stripe, and wings with some dark markings. As with *Volucella bombylans* (June - W4), their larvae live in the nests of social wasps and bumblebees. *Volucella pellucens* larvae feed on debris at the bottom of the nest. While widespread, adults prefer sheltered areas and may be seen hovering in sunlit areas. We find them within the garden where they visit a wide range of flowers.

Leucozona laternaria, (7-10mm, June to July), and *Leuconia glaucia,* (8-11mm, end-June to August)

Volucella pellucens, (10-15mm, June to July), garden

July - Week 2

Week 2

While looking for specific insects, we often notice that their foodplant is of the utmost importance. Certain insects will only be found on their host-plant and this can allow a convenient way of locating the particular creature we wish to find.

One such association, found in July, is between a mirid bug and one of the nine species of Willowherb found in Ireland. Great Willowherb is perhaps one of the most beautiful of the pink, four-petalled varieties. Tall and shrubby, it is abundant in damp locations and a welcome addition to the garden. The stem and leaves are covered in long, silky hairs, while the leaves have a further coat of shorter hair with a glistening, water-filled tip.

Great Willowherb, *Epilobium hirsutum*, (up to 2m high, flower 15-25mm, July to August)

The Willowherb name, *Epilobium hirsutum*, comes from the Greek word *epi* (upon) and *lobos* (pod). This refers to the long inferior ovary, situated below the flower, where the seeds develop. This name leads directly to its associated mirid bug, *Dichypus epilobii*, which may be found in good numbers around the leaves and stems. Once they spot you, they dash around the leaf and the merry dance for a good photograph begins.

Pale green adults with long wings (macropterous) are distinguished by the red, basal segment of their antennae, (closest to the head). They are thought to have two generations, with pale green larvae found mid-August and eggs which overwinter.

Dichypus epilobii,
(4-5mm, July to October),
on Great Willowherb, garden

July - Week 2

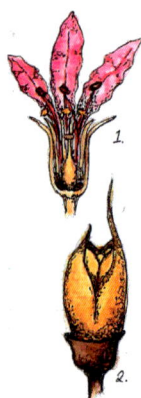

1. Flower in the longitudinal section
2. Mature seed house.

• PURPLE LOOSESTRIFE •

• *Lythrum salicaria* • Créachtach •

Purple loosestrife is generally distributed in overgrown ditches, around the margins of the bogs, and by the sides of rivers and lakes. It generally flowers from June to September and has a tall stem which can grow to almost 2m high, the purple-red flower spikes account for as much as a third of this. Its flowers are borne in dense whorls on upright spikes, each flower 10-15mm, bright reddish purple, with six crumpled, narrow petals and twelve stamens. Sometimes there can be fewer petals, different size stamens and styles. The paired, opposite, narrow spear shaped leaves are in whorls of three but those further up the stems are sometimes alternate.

The intended pollinators of the purple loosestrife are medium sized bees and certain hoverflies, with which the flowers are a great favourite. As they move between the different flowers, holding onto the long and medium stamens and styles as they probe for nectar, their bodies are dusted with pollen in three different places.

The caterpillar of the emperor moth is occasionally found feeding on the leaves of purple loosestrife, but it is more regularly grazed by four beetles.

In some places it's considered a highly invasive species and beetles that feed on the plant have been introduced in Europe to curb the invader.

Uses: The distilled juice of the plant was said to soothe sore or injured eyes, and an ointment was made from it to ease ulcers, wounds and sores. The powdered plant stops nosebleeds, and it was also prescribed for fevers, constipation, diarrhoea, and cholera. A gargle was used to treat sore throats. Chewing on branches believed to strengthen weak, bleeding gums.

Purple-loosestrife, *Lythrum salicaria*, (2m high, flower 10-15mm), project by Gráinne Lyons

Step into Nature

Purple-loosestrife, *Lythrum salicaria*, (July to August), bogs, ditches, lakes and rivers

July - Week 2

Argyresthia brockeela, (5-6mm, May to August), Birch, garden

Argyresthia goedartella, (5-6mm, June to September), Birch, garden

There are two splendid micro-moths, found enjoying the sunshine on Birch or Alder, with brassy-gold and white markings.

Argyresthia brockeela, with dark golden wings and large white spots, rests on Birch leaves, and flies in afternoon sunshine. Larvae live in Birch and Alder buds in autumn, and overwinter (fully fed) in the distorted male catkins. *Argyresthia goedartella* is more brassy, with slender semi-circular white patterns. It is found on the same foodplants, with larvae in buds or male catkins.

Step into Nature

Marsh Helleborine, *Epipactis palustris*, (50cm high, flower 17mm, July to August), boggy ground, Cloghan Lakes

Hedge Woundwort, *Stachys sylvatica*, (75cm high, flower 13-18mm, June to October), shady ditches, with Common Carder Bee

Red Bartsia, *Odontites vernus*, (50cm high, flower 8-10mm, June to September), field margins

July - Week 2

Sometimes we may find a delicate micro-moth close to rivers, lakes or ponds. The Beautiful China-mark, (a large micro-moth), is so well-named, with crisp white wings and shimmering bronze, it flies for most of the summer. They differ from other moths we have encountered in that their larvae (20mm) are wholly aquatic, living and feeding under water. Mined cases are constructed where they will overwinter. Their pupae are aquatic with sac-like cocoons (of bits of leaves), fixed to the underside of the Yellow Water-lily leaves or within the stems of Bur-reeds. Adults are subsequently found flying from dusk, close to water and their larval plant.

Beautiful China-mark, *Nymphula nitidulata*, (wingspan 16-25mm, May to September), Shannonbridge

Step into Nature

Similarly, the larger Brown China-mark (a micro-moth), is also found by ponds and lakes. They are a variable species, predominantly brown with contrasting white markings.

Brown China-mark leafmine, *Elophila nymphaeata*, (August to June), on Pondweed at Finnamore Lakes, and a larval pouch, both August

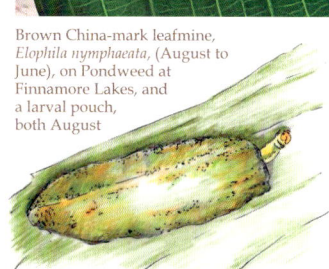

For three days, the small larvae mine around a central leaf vein of water plants - Yellow Water-lily, Pondweed, Frog-bit, Bur-reed or Duckweed. Mining results in discoloured patches and later oval cut-outs along leaf edges. This is when we may become aware of them in our pond. Larvae make a pouch in the same manner as the Beautiful China-mark, which can float independently in the water, if required.

Brown China-mark, *Elophila nymphaeata*, (wingspan 22-33mm, June to August), garden

July - Week 2

The Forest or Red-legged Shieldbug is a similar brown colour to the Spiked Shieldbug (May - W4) but the shoulders are noticeably squared off. It has red legs and a distinctive orange-tipped scutellum.

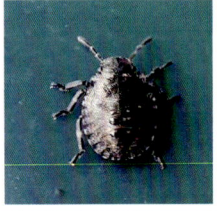

Adults may be found basking in the sunshine, on low-hanging leaves in woodland, parks and gardens. Oak is said to be their main host plant along with Alder and we have found them annually in the garden on Field Maple. Eggs are laid in batches on leaves in August, hatching by September. At risk from foraging Coal Tits, the hardy nymphs (3-4mm) overwinter on trees and we found an early instar resting on the post box, under a large Ash tree, in early January.

Stunning to look at when not trying to bite your arm, the Twin-lobed Deerfly, is found near water or damp ground. It has mesmerising metallic eyes and striking wings with brown markings, held in a triangular shape at rest.

Forest Shieldbug, *Pentatoma rufipes*, (11-14mm, July to November), early and final instar, adult, garden

Step into Nature

Twin-lobed Deerfly, *Chrysops relictus*, (8-12mm, June to August), Turraun

235

Males collect pollen and females require the blood of grazing animals to produce large batches of eggs

July - Week 3

Week 3

One of the most eye-catching garden moths is the Bordered Beauty. With shades of rich pink, orange and brown, it is unforgettable, although slightly autumnal (far too early) in colour. Females may be seen in the evening, in damp woods, laying pale, yellow eggs on Willow leaves. The eggs overwinter, turning red with white-dots and the brown-green caterpillars appear the following spring, pupating in leaves.

Bordered Beauty, *Epione repandaria*, (13-16mm, July to September), garden

Step into Nature

The rich Purple Clay is also found in the garden, sporting a range of colours from purple to red to brown with an enviable sheen. It overwinters close to the ground, as a small red-brown caterpillar with black dots and hints of yellow (August to May). For safety, it feeds at night on a range of low herbaceous plants and hides out of sight in leaf litter during the day. In springtime, it feeds on buds and young leaves of Birch saplings.

Nemophora minimella is an impressive, metallic micro-moth with incredibly long antennae. These are held aloft as it nobly rests on a sunny afternoon. Larvae feed on the seeds, then lower leaves of Devil's-bit Scabious.

Purple Clay, *Diarsia brunnea*, (16-20mm, June to August), garden

Nemophora minimella, (5-7mm, July to August), Turraun Bog

July – Week 3

Sometimes, for no apparent reason, certain species elude us. This was the case with the Small Chocolate-tip moth, at home in open, damp places, lakes, woodland and heathland. Adults are double-brooded and found in neighbouring counties, so there was no reason why they should not be in Offaly. An alternative plan is to look instead for the caterpillars. They hide by day in a larval retreat of spun leaves (usually Eared Willow), feeding under safety of darkness. Not easy to spot as the spun leaves can look like a single leaf. However, this one was discovered by the tell-tale signs of frass (caterpillar poop), small brown-black blobs on the leaf below. Strange as it may seem, this is a great way of finding all sorts of caterpillars. What a delight to find this dark caterpillar with its broad, yellow dorsal stripe and tufts of white hair, first one at Turraun and then at Killaun Bog.

Next on the wish list will be to find the rarely seen adult, which sometimes flies by day. Then we must look for irregular lines of pale, olive-green eggs on leaves. There is always something new and exciting to find.

Sketch of Small Chocolate-tip, *Clostera pigra*, (11-14mm, April to May and June to August)

Step into Nature

239 Small Chocolate-tip caterpillar, *Clostera pigra*, (May to July and September to October), on Willow, Turraun Bog

July - Week 3

Another interesting association is between Hemp-agrimony and the Plume moth, *Adaina microdactyla*. The plant grows in damp ditches, bogs and riverside areas. Cluster of pink infused flowers top the many branches.

Looking closely at stems below the flowers, we may see feeding signs (slightly bulging gall) and an exit hole left by the larvae of the plume moth. They have two broods a year with adults flying May to June and August.

Hemp-agrimony, *Eupatorium cannabinum*, (1m, July to October); gall in stem and exit hole

Sketch of Hemp-agrimony Plume moth, *Adaina microdactyla*, (13-17mm, May to June and August)

Step into Nature

The vibrant flowers of Yellow-wort are a cheerful sight in gravelly places such a quarries, eskers or cutaway bogs. Flowers which open for much of the day (10 to 4) have no scent, produce no nectar, and are not visited by insects. The annual plant depends on the scattering of abundant seeds for dispersal. These were formerly dried as a powder for treating 'humours'. Pairs of pale, blue-grey leaves are very characteristic, allowing identification of the plant when not flowering.

Fragrant Water Mint with whorls of flowers encircling purple stems, are popular with flies and butterflies. Great swathes of Water Mint in wet places and lake margins, provide an unforgettable olfactory experience. The somewhat similar Wild Marjoram, grows instead in limestone areas. It is also very much at home in the garden and, due to an abundance of nectar, is visited by a wide variety of insects.

Yellow-wort, *Blackstonia perfoliata*, (30cm high, flower 10-15mm, June to September), Cranberry Bog.
Water Mint, *Mentha aquatica*, (55cm high, flower 4mm, July to October), and Peacock, Pallas Lake.
Wild Marjoram, *Origanum vulgare*, (70cm high, July to September), and Green-veined White

July - Week 3

This small solitary bird is a joy to see as it scurries around the tree trunk in a spiral motion. Treecreepers have a distinctive and unusual downward curve to their slender beak. This is used when searching for insects on gnarled bark. They are similar in size to the Blue Tit, and are grey-brown with white underparts.

Resident throughout Ireland in mixed woodlands, hedgerows and gardens with Oak and Birch, they nest in bark crevices or behind Ivy. However, they seldom attract notice in gardens as they are quiet, inconspicuous bird with plumage that blends in perfectly with the background of tree bark. They rarely visit bird feeders, but it is the spiralling, creeping motion from bottom to top of the tree that is most noticeable. This is what drew our attention to one in the garden in late-July.

Treecreeper, *Certhia familiaris*, (12-14cm long, all year)

Step into Nature

The Yellow-bodied Black Fungus Gnat, is a small, short-lived fungus gnat. They have a bright yellow abdomen, smoky wings and a tiny out of proportion head with long antennae. Adults are nectar feeders, and can be found in numbers on umbellifers. Larvae feed on fungi and dead leaves in wet places.

Click Beetles (Elateridae family) are long slender beetles that whirr past with a noticeable sound and take off with a springing backward click, giving them their common name. They vary from black to the more common black and brown and the two varieties shown. *Ampedus* sp. are a beautiful shiny red and black, fond of boggy ground and umbellifers. Their larvae live in mouldy conifer and deciduous wood. *Ctenicera cuprea* have a wonderful metallic green and purple sheen. Males have large feathery antennae.

Yellow-bodied Black Fungus Gnat, *Sciara hemerobioides*, (1-6mm), Killaun Bog, July

Ampedus sp. (7.5-9mm, May to June), Cranberry Bog, and *Ctenicera cuprea*, (11-16mm, May to July), Burren

July - Week 4

Week 4

One of our smallest and deadliest fungi is *Ergot*. It develops inconspicuously in the inflorescence (flower head) of many types of grass. A black, banana-shaped mass grows out of the flower head (sclerotium stage), maturing in autumn. It then drops to the ground and overwinters, germinating in spring.

Ergot, (10mm), found in the garden and on Purple Moor-grass at Killaun Bog

As it is so poisonous, care must be taken to ensure that it does not enter the food chain via cultivated cereal grain. Effects are hallucinogenic and have a long, alarming history documented since the Middle Ages.

On a far more cheerful note, *Calocoris roseomaculatus*, a gorgeous rose-coloured mirid bug often rests on Oxeye Daisy and really could not look more striking. We have only recently found them in Offaly. They feed on both the flower and fruits of many herbaceous plants while larvae prefer Bird's-foot-trefoil and Clover. Eggs, which overwinter, are laid in plant and grass stems.

Step into Nature

Calocoris roseomaculatus, (6.5-8mm, July to August), Finnamore Lakes and Cranberry Bog, Gallen

July – Week 4

Parent Shieldbugs, early development see May– W4. Final instar family on a Birch leaf, July, garden

Step into Nature

247   Parent Shieldbugs, *Elasmucha grisea*, (7-9mm, all year), adult form just five days later, July, garden

July - Week 4

The perfectly named, large, Shark moth, can sometimes be found by day resting with its wings tightly closed along its body, forelegs outstretched and collar raised giving the appearance of a shark's dorsal fin. It feeds on garden and woodland flowers such as Honeysuckle, Red Valerian and Rhododendron.

Larval foodplants include the leaves of many of our common 'yellows' including Sow-thistles, Hawkbits and Hawkweeds. The larvae are cautious, usually feeding at night and hiding by day under lower leaves. Once the hard cocoon is formed, it overwinters underground.

One of the larger micro-moths is the iridescent Mother of Pearl. A luminous moth with a pearly sheen, it can be disturbed from patches of Nettle during the day in gardens, woodland and hedgerows. Larvae feed on Nettle and Elm, resting in rolled leaves.

Perhaps the most quirky moth of all is the Canary-shouldered Thorn, reminiscent of fluffy Easter chick decorations. It is widespread in woodlands, parks and gardens. Larval foodplants are trees including Birch, Alder, Elm and Limes and the hardy eggs overwinter on trees, pupating in plant debris below.

Step into Nature

Shark, *Cucullia umbratica*, (22-26mm, May to August), adult, garden.
A large, day-feeding Shark caterpillar (45mm, July to September), on Hawkbit

Mother of Pearl, *Patania ruralis*, (15-17mm, June to October), garden

Canary-shouldered Thorn, *Ennomos alniaria*, (16-20mm, July to October), garden

July - Week 4

To find certain moths, we must occasionally seek out their larval plant. Sometimes we may never see the adult. Fortunately with the micro-moth *Psychoides filicivora*, we may find both on the foodplant, Hart's-tongue. A distinctive, stout, fern of cool damp places - gardens, moist ditches, springs and even walls. It is unusual as, unlike most ferns, the fronds (leaves) are solid, undissected. As spores develop on the underside of the frond, look out for mines followed by feeding signs, in the form of an irregular mass of spun sporangia where the larva hides. Adults fly by day close to the foodplant, and this pair were found on the underside of a frond. Introduced to Ireland almost 120 years ago, probably on Asian ferns.

Hart's-tongue, *Asplenium scolopendrium*, (30-60cm, all year, new growth June. Spores and moths' irregular mass

Psychoides filicivora, (4-6mm, April to October), male and female, on Hart's-tongue, garden

Step into Nature

Blue Fleabane, *Erigeron acris*, (12-18mm, June to August), a firm favourite with exquisite flowers and dense, fluffy seed heads. Absent in many counties, it thrives in calcareous grasslands, including Cloghan Lakes, Cranberry Bog, Lough Boora and Tullamore bypass

July - Week 4

Cochylimorpha straminea, (7-10mm, May to September), Turraun Bog

Eristalis intricaria, (8-12mm, June to September), see September - W3

A delicate balancing act, Ichneumonid wasp ovipositing on stem

Common Knapweed, *Centaurea nigra*, (1m high, flower c.30mm, June to September), Clooneen Bog

Two predators, *Sicus ferrugineus*, (May - W1), above and Flower Crab Spider, *Misumena vatia*, (May - W1), below

Some plants act as a specific host to a particular insect, others support a large, mixed range, providing a complex associated food web. Common Knapweed is such a plant, host to over a dozen species of gall flies, small moths and numerous insects. The stem and seed head may contain several different species. The larvae of one moth, *Cochylimorpha straminea*, feed exclusively on developing seeds, within the upper part of the stem (July to August). Larvae of the Knapweed Gall-fly feed within the ovary of the plant, sometimes alongside a smaller gall-fly *Urophora quadrifasciata*, and Chalcid wasp larvae. Many are then preyed upon by other predators which, in turn are hyperparasitised by different species. As some lay eggs within the stem, seed head or visiting larvae, others seek out those eggs to become host for their own young. It is phenomenal to consider how the hyperparasite (often Ichneumonid wasps) knows where the initial parasite eggs have been laid. Chalcid wasps are the most common.

Externally, there is a prevailing sunny sound, a zizz of flies. *Eristalis intricaria*, the hoverfly (bumblebee mimic, see May - W1); an ovipositing Ichneumonid wasp. Then the fascinating line-up of the Conopid fly, *Sicus ferrugineus*, waiting for the arrival of a bee on top, and the Flower Crab Spider below with her hoverfly prey.

August

Week 1

How fabulous is the broad striking hoverfly, *Sericomyia silentis,* a large, black-and-yellow wasp mimic, (opposite page)? Widespread and found close to bogs or wetlands and the garden, adults most frequently visit thistles, Knapweed and Devil's-bit Scabious.

Sphaerophoria scripta is a totally different hoverfly - long and narrow. It is unusual as the abdomen, with broad yellow bands, is noticeably longer than the wings. This colouring varies depending on the time of year that the larvae develop. Cooler spring specimens can be darker and appear spotted, as the yellow band does not quite meet in the middle. In Ireland, they are predominantly found in open grassland (this individual was at Birr Demesne) and may all be non-resident or migratory. Adults visit umbellifers, while the aphid-feeding larvae may be found on herbaceous plants.

Sphaerophoria scripta, (5-7mm, June to August), brighter form Birr Demesne, dark form sketched above

Step into Nature

255   *Sericomyia silentis*, (10-14mm, June to August), on Devil's-bit Scabious, Finnamore Lakes

August - Week 1

An interesting gall that we may easily spot on Wild Carrot, is *Kiefferia pericarpiicola*. They target developing fruits in the flower heads and turn them into hard pouches. Initially green-yellow, by late-summer they have turned dark pink, becoming more conspicuous. Each pouch is home to a small orange larva. When mature, the larva leaves the gall to pupate in the soil, where the adult insect will spend the winter.

Time to look out for another mirid bug, *Apolygus spinolae*, (formerly known as *Lygocoris spinolai*) - oval, pale-green with sea-green markings. It may be found on a number of plants, Nettle, Creeping Thistle, Meadowsweet and Brambles. Females lay their eggs in the stem of their preferred host plant, which overwinter and hatch the following spring.

Kiefferia pericarpiicola, gall fly on Wild Carrot, *Daucus carota*, Cranberry Bog, Gallen

Apolygus spinolae, (5-6mm, June to September), on Nettle, Kingstown

Step into Nature

A definite favourite among shieldbugs has got to be the less common Tortoise Bug. It is about the same size as a Birch Shieldbug (October - W2), brown in colour with variable markings. Nymphs feed on grasses from May to August.

It is well worth checking out grass flower heads over the summer period, for the incredible range of insects. From mirid bugs and their nymphs to shieldbugs, flies and spiders, there is a huge range to be discovered when we take the time to look.

Not to forget our sawflies, if we check on the upper surface of Blackthorn (Pear or Cherry) leaves for signs of feeding, we might see the strange black slug-like larva of *Caliroa cerasi*. At this stage, they are covered in mucus before pupating in the ground, emerging a few weeks later as mostly black adults.

Tortoise Bug, *Eurygaster testudinaria*, (9-11mm, all year), Turraun Bog

Caliroa cerasi, (Adult 4-6mm, May to September), larva on Blackthorn, garden

August - Week 1

Wild Angelica is a fine, statuesque umbellifer found in damp areas, close to water. The domed heads and smooth stems with a pink-purple hue separate it from Hogweed. The flower heads produce nectar in great quantities and are visited by a large variety of insects; a great location for flies, mirids, bees, beetles and butterflies.

Wild Angelica, *Angelica sylvestris*, (up to 2m high, July to September)

Also found in wet places, Gypsywort looks similar to a Nettle, but the small white flowers encircle the joint between the leaves and the stem. A black dye was once made from the roots.

The more delicate satin beauty of Knotted Pearlwort, may be found in heathy grasslands. Tiny, upper stem leaves give the 'knotted' appearance and name.

Gypsywort, *Lycopus europaeus*, (20-80cm high, flowers 3-5mm, July to August)

Step into Nature

259

Knotted Pearlwort, *Sagina nodosa*, (12cm high, flower 10mm, July to September), Lough Boora

August - Week 1

Ear Moth, *Amphipoea oculea agg.*, (12-15mm, July to September)

We could undoubtedly fill each week with a wonderful array of moths, so selecting just a few is difficult. Notable varieties include the Ear Moth with broad, slightly pointed, richly coloured forewings. Larvae feed on the stems of Purple Moor-grass and Common Cottongrass. Haworth's Minor, found at Killaun Bog, has very intricate markings, larvae also feed on Common Cottongrass. Grey Scalloped Bar moths are a distinctive speckled white or pale grey. Males may be seen basking on warm ground by day or on Heather, hanging in the light before dispersing into the night. But the star of the week is the quirky red-brown Gold Spot, sporting a tufty headdress and gold-dusted forewings with silver-white marks.

Haworth's Minor, *Celaena haworthii*, (10-14mm, August to September)

Grey Scalloped Bar, *Dyscia fagaria*, (15-21mm, May to August), Clooneen Bog

Step into Nature

261 Gold Spot, *Plusia festucae,* (14-19mm, May to September), garden

August - Week 2

Week 2

There are few more breath-taking sights than the powerful, sweeping flight of a Silver-washed Fritillary. As our largest butterfly, they seldom go unnoticed when they glide past at eye level. Unfortunately, like our Hawkers, they seem to really enjoy flying with broad strokes, up and down, zooming past and rarely resting for close examination. Their preferred habitat is mixed, light woodland with sunny areas and this is where we see males as they search for emerging females. They will be found, and breed close to violets, particularly Common Dog-violet on slightly damp ground.

Males have almost 'H-shaped' marks on their wings, from which they waft scent towards the female. Females lay eggs close to their larval foodplant (Violets) but off the ground, presumably to ensure their survival during a wet winter. The ribbed, cream, pointed eggs

Silver-washed Fritillary, *Argynnis paphia*, male, (70-80mm, July to August)

Step into Nature

are laid on the trunks of mossy trees, with violets beneath. Two weeks later, as the eggs turn white, they hatch. Caterpillars overwinter on the tree trunk. By the following May, the dark caterpillars with orange spines and two narrow yellow dorsal stripes have moved to the violets to feed. They continue to feed, moult and bask in any available sunshine until they are ready to pupate from late-May through the summer. Brown pupae are suspended from vegetation, emerging as adults two or three weeks later.

As adults, the stunning orange wings are equalled by the beauty of their underwings. Delicate pink suffused with a pale wash of green, blue and yellow allows them perfect camouflage on a sunny woodland floor. Male and female adults feed on Brambles, thistles and other nectar-rich flowers, sometimes visiting gardens.

Silver-washed Fritillary, underside of wings, above, larva

August - Week 2

On a very windy morning at the Cranberry Bog, we noticed this impressive bee, fast asleep, holding on to the grass by its mandibles. One of the *Coelioxys* or Sharp-tail Bees, so named because of the pointed female abdomen (blunter in males). As with Nomada bees (April - W2) they are kleptoparasites (taking food from another). Females use the pointed abdomen to cut into the host's cell (*Megachile* or leafcutter bees, August - W2), where they lay an egg. The host egg or grub is later killed by *Coelioxys* larvae using their long curved jaws.

There are two Irish species *C. elongata* and *C. inermis*, both medium sized bees, black with white stripes and white hairs to their head, thorax and even to their eyes. They are found close to their host species by the coast, heathland, woods and gardens.

Sharp-tail Bee, *Coelioxys* sp., (6-9mm, June to August), Cranberry Bog, Gallen

Leafhoppers are a mixed group of plant-feeding bugs, some of them are monophagous (feed on single type of plant) and richly coloured and these we can identify, others are more difficult. They generally overwinter as either an egg or a nymph. One of the most striking is the vivid *Evacanthus interruptus*. A large, bright yellow and black leafhopper with variable patterning. Males' wings reach the end of their abdomen, females' wings are shorter. While noted as fairly common, we have only found one, in Clara. The similar sized *Cicadella viridis* is widespread in more wetland areas. Females have blue-green wings, males dark blue-purple, both have a yellow head with two black spots. *Zonocyba bifasciata* is another distinctive, much smaller species, yellow or white with broad dark bands. It is found on Elm or Hornbeam.

Leafhopper, *Evacanthus interruptus*, (6-7mm, June to October), grasslands, Clara

Cicadella viridis, (6-8mm, July to October) and *Zonocyba bifasciata*, (3mm, July to October) on Elm

August - Week 2

Robin's Pincushion on wild rose, by the gall wasp, *Diplolepis rosae*, Killyon.. Cross-section showing multiple chambers, Cranberry Bog

Alder Tongue, *Taphrina alni* fungus on female Alder catkin, Finnamore Lakes

Iteomyia major, gall midge on Willow, Cranberry Bog, Gallen

Pemphigus bursarius, an aphid gall on Black Poplar, Finnamore Lakes

Step into Nature

Robin's Pincushion forms on wild roses in spring, as the gall wasp, *Diplolepis rosae* lays eggs in a bud. A woody gall of many chambers develops with an odd outer coating of branched red hairs. Larvae stay in the gall, which turns brown by autumn, losing its hair. Adult wasps emerge the following spring and the cycle starts again. Another madcap gall, this time caused by a fungus, is known as Alder Tongue. After infecting the scales of female catkins, the ascomycete (sac) fungus *Taphrina alni* develops. Initially cream then red, the extraordinary growth scatters far-reaching spores by the end of summer.

There are many galls on Willows and the fused group shown here are caused by the gall midge *Iteomyia major*. Noticeable on both sides of the leaf, with openings on the underside, each gall houses a single larva. This reaches maturity by late-summer, falls to the ground where it overwinters until the following spring. One to look out for on Black Poplar is a swelling at the petiole of the leaf. It is caused by a wingless female aphid *Pemphigus bursarius*, hatched from an egg on the bark which feeds on the sap of the petiole. In a defensive attempt, the Black Poplar mass-produces plant cells which surround the aphid. Her winged female offspring develop inside the gall, and forming a beak shaped exit, break out in the summer.

August - Week 2

Scarce Blue-tailed Damselfly,
Ischnura pumilio, (26-31mm, May to
August), Finnamore Lakes

Step into Nature

Back in May - W2, we looked at the lifecycle of the Damselfly and Dragonfly. This week, we observed a recently emerged Scarce Blue-tailed Damselfly. The larvae crawled from the lake, up surrounding vegetation and the teneral adult then emerged. Pale, delicate adults have a glossy sheen during this teneral stage and take a few days to fully mature.

This is the smallest Irish damselfly, found in heaths, bogs, quarries and lakes with soft-stemmed plants for females laying their eggs. Males have a blue thorax, yellow beneath and a dark abdomen. They differ from the similar Blue-tailed Damselfly as their blue tail-light is closer to the base and has two small black dots. Immature females are initially orange, becoming a greenish-brown as they mature. They have a bright green thorax and a dark abdomen with no blue tail.

Scarce Blue-tailed Damselfly, recently emerged with exuvia at top of stem; sketch of male

August - Week 2

Carl Axel Magnus Lindman (1856-1928) was a Swedish botanist and a wonderful botanical artist, who published *Bilder ur Nordens Flora*, 1901-1905. His drawings covered all aspects of each plant's life cycle, from seed to flower to fruit. A great way to understand each stage is to try to replicate the drawings, such as this one of Trifid bur-marigold. It grows by streams, lakes, canals and boggy places. It is a gangly plant with flower heads made up of a group of tubular yellow florets.

Autumn Gentian is a beautiful flower, similar to the four-petalled Field Gentian. They are told apart by the calyx, found at the base of the flower - five in Autumn Gentian, four in Field Gentian, two small and two larger. Found in gravelly places, the flowers can have four or five petals, with a purple-blue hue and a charming inner fringe of frilly hairs. These prevent insects from entering the tube, allowing only butterflies access to their nectar. We can now start to watch out for Soapwort, introduced to Ireland for its soap-making qualities.

Trifid Bur-marigold, *Bidens tripartite*, (60cm high, flower 10-20mm. July to October)

Step into Nature

Autumn Gentian, *Gentianella amarella*, (c,25cm high, flower 13mm, June to September), Cranberry Bog, Gallen

Autumn Gentian (left), five even calyx, Field Gentian (right), *Gentianella campestris*, two large calyx enclosing two smaller

Soapwort, *Saponaria officinalis*, (up to 70cm high, flower 30mm, June to August), in bright, occasional roadside clumps

Five-petals either double form (left) and more natural separate petals (right), found on either side of the same road

August - Week 2

Megachile willughbiella, (8-10mm, June to August), Burren National Park

Step into Nature

We have five different species of wonderful Leaf-cutter bees (*Megachile*) in Ireland. They are one of our most interesting groups and, while found in many habitats, are scare. It is a true gift of nature to see the dextrous females at work. We were so fortunate to see a female *Megachile willughbiella* as she cut out circular pieces of a leaf for her nest. While she cut a section of the leaf, she rolled it between her legs until free, then flew them back in her jaws, to line the cells in her nest. Sometimes she started cutting a section, found for some unknown reason that it was not quite right, and moved on to another leaf or petal.

They make use of suitable cavities in dead wood, window frames, walls, plant pots, or bee hotels. Each plugged cell has the appearance of a cigar roll of wrapped leaves or petals, housing the egg and future larva. The neat, semi-circular cut-outs are often found in gardens or hedgerows, indicating the presence of a nest nearby.

In flight, *Megachile* females may be distinguished from other bees, by the bright orange pollen brush under their abdomen. They are considered a possible host of one of Ireland's two recorded Sharp-tail Bees, *Coelioxys* sp. which we found fast asleep at Killaun Bog (August - W2).

August - Week 3

Common Green Grasshopper,
Omocestus viridulus, (14-23mm,
April to October), Finnamore Lakes

Step into Nature

Week 3

What a wonderful way to start a new week, with this spectacular pink grasshopper. It is a Common Green Grasshopper of the Acrididae family, with a rare genetic mutation, called erythrism. Generally, they are found in grassland, heathland and even farms with long grass. Females lay eggs close to grass roots. Nymphs (which emerge in April), take four weeks of moulting to reach maturity. This fascinating pink specimen was found resting on the seed heads of Bird's-foot-trefoil, blending in surprisingly well with the reddish colour.

Also by the lake at Finnamore, we found the elongated *Donacia* sp. beetle. It is a rich bronze colour with a reddish sheen and an intricately dimpled pronotum. It has protruding eyes, long, beaded antennae, broad shoulders and powerful legs. Noticeable on sunny afternoons, we observed this one resting on vegetation by lake margins. It feeds on the leaves of aquatic plants such as reeds. Large (15mm) larvae suck air from roots of plants via abdominal spines. Brown, tough pupae may also be found within the roots.

Donacia sp. beetle, (7-11mm, May to August), Finnamore Lake

August - Week 3

Ichneumons are narrow-waisted parasitoid wasps with long, flicking antennae. Their presence on a tree, as they scurry from leaf to leaf, is usually an indicator of their host larvae. They belong to a large group of 32 families with a vast range of species of varying size, colour and overall shape. Many can only be identified with microscopic study. Some, fortunately, can be identified in the field or by photographs, such as the impressive *Hepiopelmus variegatorius*. It is a distinctive black and yellow species which enjoys well-vegetated areas. This one, found in the garden, was in search of its host, either Buff or White Ermine caterpillars. In some species, females have long ovipositors, with which they deposit their eggs into the caterpillar. As the eggs hatch and feed, the unfortunate host dies, and the adult Ichneumon emerges. Adult Ichneumons feed on nectar from flowers or honeydew from aphids.

Hepiopelmus variegatorius,
(15mm, August to September),
on Fennel, garden

Step into Nature

Another much smaller parasitoid wasp, is *Cotesia glomeratus*. Have you ever noticed a particularly sluggish Large White Butterfly caterpillar, which seem to stay in the same place for a number of days? Over time, numerous small, yellow, fuzzy cocoons grow out of either side of its body. It has been parasitised by the wasp, which laid many eggs within the caterpillar. The eggs hatch and the larvae develop inside the caterpillar, but pupate outside of the host. Roughly two weeks later, the adults emerge from the neatly opened cocoons. Sometimes, there can be another layer to the story, where the wasps' larvae have been in turn hyperparasitised by another species of wasp. This is apparent by the opening on the cocoon, as the hyperparasites chew their way out, leaving a rough jagged hole.

Large White Butterfly, *Pieris brassicae*, caterpillar on 23 July, with attached parasitic cocoons

Cotesia glomeratus, (3mm, May to August), emerged 08 August, empty cocoons after caterpillar crawled away

August - Week 3

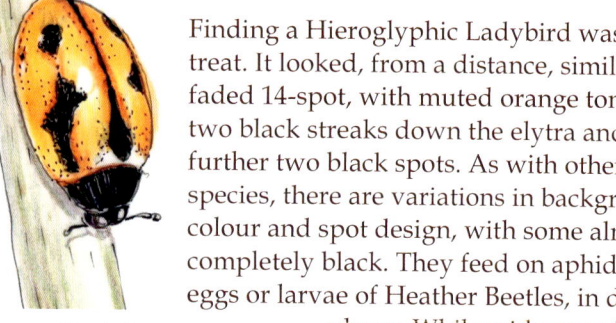

Finding a Hieroglyphic Ladybird was a real treat. It looked, from a distance, similar to a faded 14-spot, with muted orange tones, two black streaks down the elytra and a further two black spots. As with other species, there are variations in background colour and spot design, with some almost completely black. They feed on aphids and eggs or larvae of Heather Beetles, in damp places. While widespread, they are uncommon. Adults may be seen in late summer, this one was found on a small Birch sapling over Heather, at the waters edge. The 2-spot Ladybird is typically red with two black spots. But there are a number of variations including completely black with red spots. They feed on aphids found on Nettles and thistles, roses and deciduous trees, in a variety of habitats including wetlands, woods and gardens. Adults overwinter and can be located on tree trunks.

Hieroglyphic Ladybird, *Coccinella hieroglyphica*, (4-5mm, May to October), Cranberry Bog, Gallen

2-spot Ladybird, *Adalia bipunctata*, (4-5mm, May to October), Finnamore Lakes and garden

Drymus sylvaticus, (4-5mm, all year), in leaf litter, garden

Despite being a common land bug, we have only observed the metallic *Drymus sylvaticus* once, in the garden on leaf litter. Overwintering as adults, it matures fully between May and June, laying batches of 28 pink-yellow eggs in dry, mossy places.

Pithanus maerkelii, (4-5mm, June to August), garden

Some further mirid bugs we may observe include the ant-like Capsid bug, *Pithanus maerkeli*. Usually micropterous (tiny wings), but some females are fully-winged. Found in damp meadows, they lay eggs in grass stems in July, which overwinter, hatching in April. Larvae can feed on young leafhoppers. The small, delicate, speckled green *Malacocoris chlorizans*, is well worth searching for on the underside of Hazel leaves. This imposing White Plume Moth, with forewings of crisp white, tapering lobes, can be found by day in hedgerows.

Malacocoris chlorizans, (3-4mm, May to October), underside of Hazel leaf, garden

White Plume Moth, *Pterophorus pentadactyla*, (12-16mm, March to September), garden hedgerow

August - Week 3

While out on our walks at this time of year, we should always keep an eye on the leaves of Willow and Birch. We may be fortunate enough to catch sight of some more sawflies, eggs, larvae and adults. The superb blue and yellow larvae of *Euura ferruginea* are hard to miss and this one was munching on a tiny Willow sapling close to Clara boardwalk. Adults are also very striking and this large, gracious female obligingly stayed still, poised at the tip of a Willow leaf in the Cranberry Bog. Reddish-brown with a green abdomen, large antennae she has large, watchful eyes.

Euura ferruginea, late instar larva, Willow, 18 August, Clara boardwalk

Euura ferruginea, (7-9mm, May to June and July to September), adult, Willow, 25 August, Cranberry Bog, Gallen

Step into Nature

Sawflies lay their eggs in a variety of ways on different plants. *Arge ustulata* oviposit into and around the edge of a Birch leaf. Raised pockets are visible on both sides of the leaf. As the egg hatches, the early instars make their own nibbled cut-out as seen on a small Birch sapling close to the water at the Cranberry Bog. They also feed on Willows and Hawthorn. Adults have brown-tinted wings, while their body has a beautiful green metallic sheen, particularly noticeable on sunny days. A good place to find them is feeding on umbellifers such as Angelica.

Arge ustulata, early instar larva, on Birch, August, Cranberry Bog, Gallen

Arge ustulata, (7-10mm, May to July), adult, on Wild Angelica, Finnamore Lakes

August – Week 4

Week 4

A curious, fragile-looking bug, *Metatropis rufescens*, may be found on the leaves of Enchanter's-nightshade. The plant, with heart-shaped leaves, favours shady locations under trees. Dainty white flowers rise up over the leaves in tall, loose racemes. Adult bugs overwinter, laying single eggs in stem or leaves of the host plant. Three weeks later, they hatch, and over the following eight weeks, go through five instars. We should search for adults from mid-August as they feed on seeds on Enchanter's-nightshade. Their long, thin body varies from reddish-brown to black, and is raised off the ground on thready legs with swollen femoral tips. Characteristically long, thin antennae have four distinct segments, a truly remarkable insect.

Stiltbug *Metatropis rufescens*, (9-11mm, June to October), Knockbarron

Enchanter's-nightshade, *Circaea lutetiana*, (40cm high, flower 4-7mm, May to August), Knockbarron

Step into Nature

The Sputnik Spider is a tiny almost transparent spider that occurs fairly high on shrubs, garden evergreens, and foliage of the tree canopy. One of the Comb-footed spiders, *Theridiidae* family, they are more likely to be found protecting their unique egg sac. It is an extraordinary, white, spikey case, found on the leaves of Oak or Birch. Another particularly attractive spider with various colour schemes, is the Candy-striped Spider, pale yellow with a dark pink stripe, and sometimes white with rows of black dots. It is found in gardens on low vegetation, or we might see the pale blue egg sac that females neatly wrap in a rolled leaf for security.

A variable hoverfly to watch out for is the migratory *Eupeodes corollae*, with broad yellow markings.

Sputnik Spider, *Paidiscura pallens*, (1.5-2mm, male April to June, female April to November), garden

Candy-striped Spider, *Enoplognatha ovata*, (5mm), garden

Eupeodes corollae, (5-8mm, July to August), visits garden flowers

August - Week 4

Dicyphus stachydis, (4-5mm, July to September), Hedge Woundwort, garden. Slightly more oval shape adults darken as they overwinter, and lay eggs May to June

Dicyphus pallicornis, (3.5-4mm, all year), on Foxglove, garden. Two generations ensure finding adults or larvae on the underside of Foxglove leaves

Pinalitus cervinus, (4mm, all year), on Hedge Woundwort, Kinnitty Demesne. Usually found on Lime, Hazel, Ash or Ivy. Up to six eggs laid on Ash or Hazel buds

Deraeocoris lutescens, (4mm, all year), on Lime, Birr Park. A small, shiny predatory bug, lays eggs deep in woody tree growth May to June

Step into Nature

Two fine big hawkers are currently on the wing. First we look at the Common Hawkers, one of our larger, jazzy dragonflies. They can be heard taking off from sheltering Gorse, with a clickety-click of their wings. Strong and powerful fliers, they have notable yellow costa (leading edge to wing). Females are brown with yellow spotting. Males are blown-black, waisted with pale blue spotting and yellow stripes to the thorax. They enjoy a wide-ranging habitat including pools, lakes and streams.

A confusion species are the smaller, but similarly coloured, Migrant Hawkers. However, they do not have a yellow costa and male spots are a deeper blue. But their characteristic mark is the 'golf tee' below the thorax. Females are also brown with yellow spots. They enjoy ponds and lakes that are well-vegetated and not too deep.

We have been extremely privileged to spot both in the garden.

Common Hawker (top), *Aeshna juncea*, (74mm, June to October), Cranberry, Killaun and Turraun Bog, and garden

Migrant Hawker, *Aeshna mixta*, (63mm, August to November), Clooneen Bog and garden

August – Week 4

While we are never supposed to have favourites, the end of August brings two of the most beautiful flowers to our attention. Grass-of-Parnassus has a sculptural quality of the utmost delicacy. A single perfect flower rises above surrounding damp grass on a tall, slender stalk. The five satin petals have translucent veins giving them a waxy depth, and at the centre, a domed ovary. This is surrounded by the unusual modified, yellow stamens with glistening tips, so developed to attract visitors.

Grass-of-Parnassus, *Parnassia palustris*, (30cm high, flower 15-20mm, July to September), Clonfinlough

Step into Nature

The diminutive Autumn Lady's-tresses, are a rare and wonderful find in limestone grassland and perhaps the most beautiful orchid. As the scientific name would suggest, the short spikes twist in an elegant spiral presenting the unfolding flowers. The sparkling diaphanous white-green flowers produce an abundance of nectar for pollinating bumblebees. Basal leaves can take up to 11 years to develop from seeds, and as one plant flowers, an adjacent rosette develops for the following year.

This example is from the Burren but we have also found Autumn Lady's-tresses off the Ridge Road, Birr

Autumn Lady's-tresses, *Spiranthes spiralis*, (10cm high, flower c.6mm, August to September)

August - Week 4

Tenthredo colon, on Great Willowherb, one of four we found in the garden. Elongated black and red adults fly May to August

Euura papillosa, on Aspen, Kingstown, similar to *Euura Pavida* (May-W2), but they have an extra row of lateral dots

Cladius grandis, hairy, gregarious feeders on the underside of Aspen leaves, Kingstown. Yellow and blacks adults fly May to September

Diprion pini, feed gregariously and only on pine, Scots Pine, Killaun Bog. Recognisable, cream with black lateral spots and an orange head

Step into Nature

Shown here are a selection of interesting sawfly larvae to look out for, which may be found on leaves in the garden or on local walks.

When visiting the bog, we may observe some fast-flying bees at work, as they dash from one Heather plant to another. *Colletes succinctus*, from a family of Plasterer Bees, are medium-sized, furry solitary bees and the largest Irish species (2nd-largest, after the arrival of the Ivy Bee). They nest in light soil and using their tongue, line nest cells with a waterproof, fungus-resistant layer. Eggs are attached to the top of the cell and the base filled with a liquid mix of pollen and nectar.

Colletes succinctus, (6.5-8.5mm, July to October), on Heather, Cranberry Bog, Gallen

August - Week 4

Red Clover (May to October), of the pea family, has a spherical head packed with over 100 individual pink flowers. They open starting from the base of the head first. The leaves are quite beautiful, with three hairy leaflets, often decorated by a paler 'v-shaped' line. The agricultural benefit of Red Clover was acknowledged in the 19th century, when the former three-year rotation was increased to allow an additional Clover year. In a symbiotic (mutually beneficial) relationship, root nodes shelter and feed a bacterial partner which help fix nitrogen in the soil, drained by previous crops. Red Clover was and can be used to make a purported excellent wine.

While only bees can pollinate them, they are visited by other insects. One unique partnership is with the micro-moth *Coleophora deauratella*. A fascinating way to find them is at case-bearing larval stage, as larvae feed on the developing seedcases of Red Clover. They bob backwards and forwards, tucked between the flowers, some of which have already turned brown (part of their resourceful disguise). That small movement is a help when trying to find them. We found a number of them on a large clump of clover in a sunny location, the white-tipped tri-valved case (three sections) also assists in locating them.

Step into Nature

Sometimes they abseil from the flowers, dangling by a silken tread, to a leaf below. Moving with surprising speed and dexterity, they wriggle off in search of a more suitable flower head.

As adults, they are one of six green or bronze metallic species, and notoriously difficult to identify. They vary in size from 7-12mm and *Coleophora deauratella* are generally the larger at c.12mm. The bronzy colour is darker towards wing tips and their antennae are black with white tips. They fly from June to July, before noon on sunny days around grassy places, close to their food source.

Coleophora deauratella feeding on Red Clover, *Trifolium pratense*. Note the three white-tipped, trivalved cases

Coleophora deauratella, case with larva peeping out as it moves along the clover leaf. Sketch of metallic adult

September

Week 1 Throughout August, we gallantly pretend it is still summer, ignoring signs of the approaching autumn. By September, with darker evenings, cooler days with more wind and rain, we can no longer deny the season. But it is an opportunity to view nature's softening palette. As isolated Birch leaves turn golden and fall, Rowan berries provide flocks of Starlings with a rich diet. Flowering plants echo these colours, existing insects may darken and those emerging come forth in garments of golden brown with tints of brown or red.

One such autumnal moth is the Sallow, found with a moth trap in the garden and at Killaun Bog. We were also fortunate to see one during the day in the Cranberry Bog. So regal, with buff yellow wings and dark markings, allowing it to blend into the browning leaves where it rests. Overwintering eggs are laid near plant buds. The brown freckled caterpillar first feeds on Willow catkins until it falls to the ground, moving on to low-growing vegetation. It fashions an underground cocoon, pupating over six weeks, emerging as an adult.

Step into Nature

Sallow, *Cirrhia icteritia*, (14-17mm, September to October), Killaun Bog

September – Week 1

Large-flowered Evening-primrose, *Oenothera glazioviana*, (1.8m high, 6-8cm, June to September)

Traveller's-joy, *Clematis vitalba*, (Climbs up to 15m, 15-20mm, July to September), Kilcormac

Bladderwort, *Utricularia* sp., (40cm high, flower c.7mm, June to September), Pallas Lake

Common Fleabane, *Pulicaria dysenterica*, (60cm high, flower 15-30mm, July to September), Pallas Lake

The big, floppy flowers of Large-flowered Evening-primrose rise high over roadsides, quarries and even sandy soil. It is a garden escape and while seeds may last for decades in the ground, the flowers only bloom for two nights. The fragrant evening scent attracts moths, their pollinator of choice.

Traveller's-joy is a vigorous wild clematis of hedgerows. It comes from the *liana* family of woody climbing trees with rope-like stems. Initially flexible leaf-stalks wrap around higher branches of adjoining plants, firmly fixing it in place. The creamy-green flowers are pollinated by bees and flies.

Bladderworts are aquatic plants found in boggy waters or lake edges. Insects in the water are trapped in small underwater bladders that are connected to hair-like leaves. As with Sundew and Butterwort, insect nutrients make up for deficiency in the habitat, helping to sustain the plant.

Common Fleabane, despite the name, is not very common, but the large, deep-yellow flowers are easy to spot. They grow near water, rivers, canals or lakes where the slightly woolly leaves rise up through long grass. Formerly, the smoke of burning flowers was thought to ward off unwanted insects such as fleas, giving the common name.

September - Week 1

A particularly handsome, large mirid bug, *Pantilius tunicatus,* is found on Alder, Hazel and Birch. Red-green adults with speckles of black blend into the autumnal buds, but do not overwinter. Both adults and larvae feed on buds, catkins and young shoots. Small eggs are laid at the base of buds, and leaf or catkin scars. Young adults are a yellow-green colour. They are found in woodlands or gardens.

This glossy hoverfly, *Platycheirus granditarsus*, has broad orange markings to the centre of the abdomen. Males have thick, modified tarsi to front and middle legs. They favour wetland and damp grassland habitats, fens and the borders of raised bogs. A metallic, gleaming thoracic dorsum and shinny wings provide such a luxurious appearance as they rest on grass, *Juncus* (rushes) stems or umbellifers.

Pantilius tunicatus, (8-10mm, September to October), on Alder, garden

Platycheirus granditarsus, (5-8.5mm, June to September)

Step into Nature

Mellinus arvensis is a field digger wasp with a distinctive rectangular head, often found at brownfield sites. Individual nests are built in south-facing sites, with several cells in each burrow. Females prey on flies and leave a few in each cell, into which she then lays one egg. As the wasp larva hatches and develops, it feeds on the accumulated flies. The following year, they emerge as an adult.

Andrena is the largest genus in Ireland of mining bees and we have come across a few earlier in the year (April - W2). With Devil's-bit Scabious in flower, we should take a look for the Small Scabious Mining Bee. It is a scarce bee, Critically Endangered on the Irish Red List and any available records can help map its distribution. It favours an abundance of Devil's-bit Scabious, in either boggy or limestone areas.

Mellinus arvensis, (8-12mm, June to October), Kinnitty Demesne

Small Scabious Mining Bee, *Andrena marginata*, (6-9mm, August to September), Finnamore Lakes

September – Week 1

Zebra Jumping Spider, *Salticus scenicus*, (6mm), garden

Heliophanus sp., (6mm), on vegetation, Burren

Garden Spider, *Araneus diadematus*, (males 4-8mm, females 10-15mm), Finnamore Lakes

Arctosa perita, (6-9mm, February to November), Clooneen Bog

Four-spotted Orbweb Spider, *Araneus quadratus*, (12-15mm), Tullamore Bypass

Step into Nature

Some of our spiders seem more visible at this time of year, in gardens and bogs. The Zebra Jumping Spider can be found on sunny walls and we were delighted to see this one enjoying the sunshine in the garden. It does not build a web to catch prey, but instead uses its large eyes to watch out for them. Then, calculating the distance it jumps, catching its prey with strong front legs. There are other species of Jumping Spider and *Heliophanus* sp. is such a great example, silky black with smartly contrasting yellow-green legs and palps. Garden Spiders can be seen in hedgerows and gardens from late summer to October. The abdomen varies from black or brown to orange, with a distinctive central, white dotted cross. It makes orb webs a metre or higher from the ground, on fences or tree branches, to catch a broad range of flying insects. Eggs overwinter in silken bags within tree bark and tiny spiderlings emerge the following spring. The Four-spotted Orbweb Spider is large with a bulbous abdomen of variable colour and four white spots. It is found in bogs or grassland with wide webs linking strands of tall grass.

The eye-catching, patterned *Arctosa perita*, found in coastal situations, also favours bare areas in heathland. Females are thought to overwinter. They may be seen scurrying across bare patches of bog and retreat into well-concealed burrows when alarmed.

September - Week 1

The latest and smallest darter is the Black Darter, abundant by acid or nutrient deficient lakes, bogs and heaths. Mature males are very distinctive as they are the only Irish black dragonfly. Females are golden-yellow with black underside, (both have black legs). Easily disturbed, they are frequently found enjoying the sunshine on vegetation or a boardwalk. Also soaking up the warmth of the boardwalk, is the Large Marsh Grasshopper. It is such an impressive insect, of wet, acid bogs, such as Mongan and Clara. Males are very active on sunny days and can fly up to 10m. The larger females hide in vegetation and are our largest grasshopper. A smooth olive-green with keels that are almost straight. They are quite remarkable with powerful rear thighs with a red underside reaching black knees and bright yellow tibiae that are clad in black spines. As with all our grasshoppers, they feed on plants, lay eggs in grass roots in autumn and overwinter as a nymph.

Large Marsh Grasshopper, *Stethophyma grossum*, (21-36mm, August to September), Clara boardwalk

Black Darter, *Sympetrum danae*, (32mm, July to October)

Step into Nature

Similar to very small grasshoppers, *Tetrigidae* are a large family of squat, non-singing Groundhoppers found in damp places with bare ground. The Slender Groundhopper is a grey-brown groundhopper with a distinctive prothorax extending beyond the abdomen.

Slender Groundhopper, *Tetrix subulata*, (9-12mm, March to June, September to October)

A slightly smaller species is the Common Groundhopper. It has a prominent crested ridge which does not extend past the hind femur. Both species are well able to swim to safety, if they land in a bog pool or puddle.

Immature Common Groundhopper, *Tetrix undulata*, (8-11mm, April to September), Cranberry Bog

September - Week 2

Week 2

There are still many mirid bugs about, but some are more problematic than others. Species in the genus *Lygus* are difficult to identify, and this shiny example may be *Lygus wagneri*. Adults mature by August, overwinter and oviposit between May and June. They favour hedgerows with Dock, Nettle and St John's-wort.

An easier species to identify is the larger, Lucerne Plant Bug, *Adelphocoris lineolatus*. It is frequently seen poised on flowering plants in the sunshine. Both adults and larvae feed on leaves, stems and flowers of the pea family, Vetch and Clovers. Around 48 eggs are laid in plant stems from mid-July. Young, dark green larvae with thick antennae become adults from July onwards.

Lygus wagneri (5-6.5mm, all year), on Oxeye Daisy, Finnamore Lakes

Lucerne Plant Bug, *Adelphocoris lineolatu*, (7-9mm, July to October), Kinnitty

They are of similar colour and proportion to the Potato Capsid Bug, which is common in gardens and hedgerows. It feeds on new growth of a range of plants, laying eggs in stems.

Pale green mirids, *Plagiognathus chrysanthemi* dash around the stems of Yarrow, Black Medic or grasses. Speed and inconspicuous colouring can make them difficult to spot in dry waste areas, but they are usually around in large numbers. Overwintering eggs hatch in May and larvae mature by late June. Females oviposit in plant stems between July and October. They can also be found on Nettles along with the variable and common, *Plagiognathus arbustorum*. It has a similar life cycle with adults appearing at the start of July.

Potato Capsid Bug, *Closterotomus norwegicus*, (6-8mm, May to October), *Rudbeckia*, garden
Plagiognathus chrysanthemi, (3.5mm, June to October), Ballylin Bog
Plagiognathus arbustorum, (4mm, July to October), garden

September - Week 2

Due to the wonderful selection of bugs and galls, we have paid scant attention to the many flowering plants on our walks. Perforate St John's-wort is one of several Hypericum species. Tiny black dots on the flowers and leaves help to separate this from other varieties. But it is the leaves that give it the common name - pinhole perforations are noticeable when held up to the light. In London, their flowers were gathered for bonfires on the Eve of St John, and in many countries were used as a charm to ward off evil spirits. Always a delight to come across, Fairy Flax is a delicate little plant that rises up on hairless stems in wet or dry grassland. The five, pointed-petaled white flowers have transparent veins with a yellow centre and are encased in five beaded sepals. If the yellow centre fails to attract insects, self-fertilisation will take place as the stamens bend towards the centre of the flower.

Perforate St John's-wort, *Hypericum perforatum*, (70cm high, flower 25mm, May to September), Cranberry Bog

Step into Nature

Fairy Flax, *Linum catharticum*, (20cm high, flower 5mm, May to September)

September - Week 2

Hartigiola annulipes, (5mm), gall midge on Beech leaves, garden

Fig Gall, *Tetraneura ulmi*, (15mm), aphid gall on Elm, Birr

A gall midge, *Hartigiola annulipes*, creates these raised growths on Beech leaves. In the spring a small dome is visible on the upper and lower leaf surface. By autumn, it has elongated into a slightly hairy cylinder, turning yellow-green and then brown. The gall falls to the ground as its single white larvae reaches maturity. If we have missed this stage, the remaining oval cut-out on the leaf is another indication. Larvae pupate in ground cover and by the following spring, adults fly up to fresh leaves to lay their eggs.

Elm trees are host to the upright, aptly named Fig Gall found on the upper surface of the leaf.

They are caused by the aphid *Tetraneura ulmi*, and there are several in each of the smooth pouches. During the summer, mature aphids (asexual females) fly from a small opening in the gall to breed at the roots of grass. They fly back to Elm trees in the autumn giving birth to tiny wingless aphids. These then mate and females lay one egg which hatches in spring (the asexual female). Feeding on the underside of the leaf causes first, a slight swelling to develop on the opposite side. When large enough, the aphid moves into the developing gall to produce their own offspring. Up to seven can form on one leaf.

Have you ever noticed the tip of a Bracken frond tightly rolled-up? It may be caused by the larvae of a gall fly, *Chirosia grossicauda*. In late-summer, it tunnels into central veins, forcing the tip to roll backwards. It pupates as the Bracken withers, overwintering in the leaf litter below.

Chirosia grossicauda, (August to October), gall fly on Bracken, Killaun Bog

September - Week 2

There is something very endearing about the small, sawfly larva, *Platycampus luridiventris*. It hugs closely to the underside of Alder leaves, usually against a vein or mid-rib often remaining unnoticed. It is pale green, very lightly speckled and only once have we clearly seen the eyes as it began to lift its head. Developing slowly from August into October, it leaves small oval holes while feeding on Alder. Adults are generally black with red-yellow legs. Despite seeing many larvae in the garden, we have yet to encounter an adult. So while it is really satisfactory finding all stages of the sawfly lifestyle, we are not always that fortunate. But it does provide us with lots to look out for in the future.

We have only found the attractive adult *Eutomostethus ephippium,* and not the grass-feeding larvae. Similar to the larger, black and red adult *Eriocampa ovata* (June - W2), but the thoracic red extends underneath. There is also an all-black form.

On other occasions, sawflies are more easily found as a leafmine such as *Fenusella nana.* By checking Birch leaves, we can spot these mines as they start at the tip of a vein by the leaf edge, moving inwards to form a large blotch. Frass is used to plug the opening, and the small, single green larva matures to a dark yellow.

Step into Nature

Platycampus luridiventris, (5-6mm, May to June), underside of Alder leaves, garden

Eutomostethus ephippium, (4-5mm, May to September), garden

Some larvae are a treat to find and a specific plant certainly helps. The sawfly (Raspberry Sawfly), *Cladius brullei*, is initially pale green with a black head, becoming a dramatic, dark grey-black, with a pale underside during its final instar. A very dapper larva. Despite flying from May to September, we have not yet found the dark adult (4.5-7mm).

Fenusella nana, (summer and autumn), on Birch, garden

Cladius brullei, on Raspberry, garden, September

September - Week 3

Week 3

As we move into the third week of September, we must make the most of any snatches of sunshine. Many flowers are still in bloom such as Common Knapweed, Great Willowherb, Common Centaury, Blue Fleabane, Water Mint, Yarrow and Common Ragwort. We were very fortunate to get one such sunny afternoon at Derryounce Lake. On that particular day, we spotted a flash of yellow, flying low to the ground from one Red Clover to another. It took persistent chasing, creeping up to the flower to catch a photograph of our first (and so far, only) exquisite Clouded Yellow. Full of exceptional charm, they are a delight to behold.

Clouded Yellow butterflies are erratic migrants from southern Europe and northern Africa. More of a south coast species, they are sporadically found across Ireland in Clover filled grassland. They are a subtle wash of green and yellow with prominent spotting in black, brown and ringed white.

Females lay a single gridded egg on Clover or Bird's-foot-trefoil. Just a week later, a green larva emerges which develops a yellow lateral stripe with pink and white dashing, taking 5-6 weeks to mature. The 20mm yellow-green pupa is fixed to a plant stem and the adult emerges about two weeks later.

Step into Nature

Clouded Yellow, *Colias croceus*, (58mm, April to October), on Red Clover, Derryounce Lake

September - Week 3

The Vestal, *Rhodometra sacraria*, (12-14mm, August to September), garden

We also have some marvellous migrant moths and an unusually warm, cloudy September evening with little wind makes perfect conditions for moth trapping.

One such surprisingly delicate species, The Vestal, visited the garden from southern Europe and North Africa. While more common in coastal locations, they occasionally make it to the Midlands, (as with the Clouded Yellow). This was only the second Offaly record of this scarce autumnal migrant.

Another migrant on the same evening was the micro-moth, Rusty-dot Pearl of the *Udea* family.

Rusty-dot Pearl, *Udea ferrugalis*, (9-11mm, late-summer to autumn), garden

Step into Nature

An adult Vapourer moth was also a first at Derryounce Lake, resting by day high on a Birch tree. Males are very distinctive, orange-brown wings with a conspicuous white spot. They overwinter as a huge batch of several hundred eggs, outside the empty cocoon of the flightless, sedentary female. The more familiar hairy larvae hatch from May to early September, and feed on a variety of broadleaved trees. We have found several larvae in the garden but no adults so far.

Vapourer, *Orgyia antiqua*, (12-17mm, July to October), Derryounce Lake. Caterpillar, garden

The Black Rustic is a sumptuous moth with dark glossy wings and cream edged kidney mark. Adults feed on Ivy flowers and blackberries. Larvae overwinter from October to May, feeding on Blackthorn, Heather and Clover by night.

Black Rustic, *Aporophyla nigra*, (17-21mm, September to October), garden

September - Week 3

Adelphocoris seticornis, (6-7mm, July to September), Charleville Demesne

Eristalis intricaria, (8-12mm, June to September), garden

Field Grasshopper, *Chorthippus brunneus*, (13-24mm, July to October), Killaun boardwalk

Step into Nature

Adelphocoris seticornis is a most attractive mirid and we were delighted to find this one, a county first, at Charleville Demesne. They prefer damp or marshy places with long grass or rushes and Vetch, feeding on unripe fruits. As with many mirids, eggs overwinter near the top of plant stems, and the dark larva with yellow markings emerge the following spring. Adults develop from July until September. We still have hoverflies to watch for including *Eristalis intricaria*, (with characteristic wing loop, April - W4). Another bumblebee mimic, particularly females with their white 'tails'. They prefer damp places with low-growing blue and purple flowers, notably thistles.

Our last grasshopper of the year is the very common, Field Grasshopper which varies in colour from green, yellow, brown, red, grey or black. Grasshoppers are predominantly plant-feeders biting off pieces with their mouth parts. They can walk, jump and most can fly. Field Grasshoppers are one of the best fliers, and very active on a sunny day, travelling several meters. Yet most of the time they walk, and nymphs are found close to where they hatched, (bare ground or ant hills), from May to July. As they grow and require more of their food source, they move to areas of taller grass nearby.

September - Week 3

Pea Gall, *Cynips divisa*, (5-7mm, late summer to autumn), gall wasp, Oak, garden

Smooth Spangle Gall, *Neuroterus albipes*, (c.5mm, July to September), gall wasp, Oak, garden

Oak trees support a vast range of flora and fauna including many specific galls. Pea Galls form on the underside of Oak leaves in late summer. Their colour changes from cream to pink, red and brown. By autumn, the galls fall off the leaves to the ground. In winter, the adult wasp exits the gall from a central circular chamber, flies up and lays eggs in dormant buds.

Another prolific family of gall wasps are the *Neuroterus* spangle galls, including the Smooth Spangle Gall. Also on the leaf underside, these thin, hairless disc galls can be cream or pale green with a raised rim, sometimes sporting a pink trim.

Step into Nature

Mature by autumn, the galls fall into leaf litter below and it takes until the following spring for the adult wasp to emerge.

A particularly attractive variety in the same family, is the Silk Button Gall, which look like pom-poms made of silky golden thread. By late summer, the underside of Oak leaves can hold hundreds of these small discs with a sunken centre. They mature in the autumn and one adult wasp emerges from each gall in the spring. The Common Spangle Gall can be found anywhere on the Oak leaf. A flat yellow-green disc with no rim and a raised central mound, the gall is covered with tufts of hairs, red or orange-brown. The galls fall from the leaf and overwinter in the leaf litter. A single adult emerges in spring and lays eggs in Oak buds.

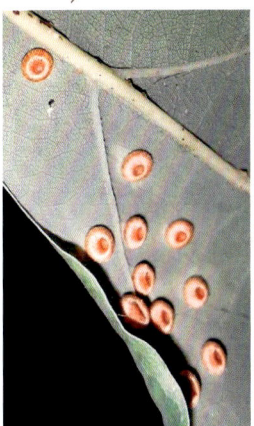

Silk Button Gall, *Neuroterus numismalis*, (5mm, August to October), gall wasp, Oak, garden

Common Spangle Gall, *Neuroterus quercusbaccarum*, (c.5mm July to October), gall wasp, Oak, garden

September - Week 3

Now we need to keep an eye out for mushrooms. Puffballs are an emerging species which, as the name suggests, form a rounded spore-filled sac and look similar to a hot air balloon. The pear-shaped Common Puffball is found in woodland and gardens. It is round-headed with a short, broad stalk. Initially pinky-white and covered with conical spines which gradually fall off, leaving a dimpled surface. When mature, the apical pore (tip) opens releasing the spores. A gentle prod with a stick produces this very dramatic cloud (or puff) of spores as they exit the fruit body and float away on the wind. Sometimes we find the darker brown, mature form after spores have been released and the sac starts to collapse.

Stinkhorns employ a different technique to disperse their pores. The thick stem rises up to a bell-shaped, fruiting cap. A honeycomb surface is covered in a brown fetid slime with a pungent smell of faeces or rotten meat.

Common Puffball, *Lycoperdon perlatum*, (c.7cm high, 1.5-6cm wide, late summer to autumn), garden

Irresistible to flies who travel long distances to land on the sticky surface, and then bring the spores with them as they fly off again. The stalk grows out of an initial 'egg' from leaf litter close to rotting tree stumps, in gardens or woodlands.

Many invertebrates are attracted to particular fungi either as a source of food (flesh or spores) or as a place to lay their eggs. Some use fungi with a broad cap as a refuge, which in turn leads to visiting predators. Specific flies attracted to Stinkhorns include the Flesh-fly, Bluebottle, Greenbottle and Cluster-fly.

Stinkhorn, *Phallus impudicus*, (up to 20cm high), covered with flies, Knockbarron

Stinkhorn with slime removed by flies, revealing honeycomb or 'morel-like' surface

September - Week 3

Fairy Inkcap, *Corpinellus disseminatus*, (c.1.5cm high), garden

Fairy Inkcap is of the same family as Glistening Inkcap (January - W1) but is much smaller, white, and grows in dense, spectacular clusters on deciduous tree stumps.

Also growing in groups on trees are the familiar yellow, Sulphur Tufts. The stipe (stem or stalk) is hollow and often has to curve as one jostles against another for space.

Green Elfcup is an unusual type of fungus, sometimes easier to discern during the autumn when in mycelium form.

Step into Nature

It grows through fallen branches, particularly Oak, staining them a dramatic blue-green. Less often, we may catch sight of the small green, cup-shaped fruiting bodies. The green-stained Oak was used as a decorative furniture inlay, known as Tunbridge ware in the 18th and 19th centuries.

Sulphur Tuft, *Hypholoma fasciculare*, (c.9cm high), garden

Many fungi are difficult to identify with certainty, due to their changing appearance within a matter of days. One such example is the Parasol family. A very attractive species of woodland and parks, with a stout stipe and, when fully grown, a wide cap with shaggy scales.

Green Elfcup, *Chlorociboria aeruginascens*, (1-5mm), Kinnitty

Three images of possible Shaggy Parasol, *Chlorophyllum rhacodes*, (stipe 15cm, cap 18cm)

Initial globular form and four days later with broad cap, gills and ringed stipe, garden

September – Week 4

Week 4

We were very fortunate, when out late in the garden, to see the swooping form of an elusive Barn Owl. It landed on a garden Oak, checked the surrounding area then took off, in total silence, flying low to another tree. The flash of brilliant white as it flew by with heightened quietude, confirmed the species.

Barn Owls are one of our Red-listed species, due to a significant decline. They are now a scarce resident, found mainly in the centre and south of Ireland. For this reason, intensive and thankfully successful efforts are being made by BirdWatch Ireland to monitor and protect their nesting sites. They often breed in ruined stone buildings, castles and sometimes outbuildings. Purpose built nest boxes are also now used. These beautiful birds are rarely seen during the day, and hunt by night for small mammals and frogs. A viable breeding site will be located where large quantities of prey can be found.

Barn Owl,
Tyto alba, (32-36cm)

Step into Nature

Grey Wagtail, *Motacilla cinerea*, (18-20cm)

The Grey Wagtail, a common Irish resident, is always busy, zipping around the edge of the garden pond with its synonymous jerky, aptly-named wagtail movement. They are a similar size to the Pied Wagtail but have a longer tail and vibrant lemon-yellow underparts. Associated with fast-flowing rivers, they feed on insects, and nest close to water, in wall crevices, buildings or under bridges.

Dippers are also widespread residents, waterside birds of fast flowing rivers. Rich, dark brown, plump birds with a brilliant white throat and breast. They bob along in shallow sections of the river, perching on rocks or flying low over the water. Hunting underwater for caddisfly larva or mayfly nymphs, they can swim and walk on the riverbed. Frequently seen by the Camcor River in Birr, look out for one of their perching boulders with tell-tale bird droppings, used for diving into the water for insects. They build domed nests in holes along the river bank, under bridges, or successful nest boxes. Juveniles disperse in search of suitable habitats. It was such a wonderful surprise when a young Dipper spent a morning in the garden checking out our pond.

Dipper, *Cinclus cinclus*, (17-19cm), the Irish Dipper is the only European aquatic songbird

September – Week 4, some leafmining moths

Hazel: *Phyllonorycter coryli*, (July and September to October). This is an easy leafmine to find, on the upper side of Hazel leaves, over veins. The mines are in the form of silvery blobs. As they develop, they contract, causing the leaf to fold upwards, garden

Hazel: *Parornix devoniella*, (July and September). This is another straightforward example. The early mine is in the form of a square or triangular blotch with brownish lower epidermis. Later, it moves to the leaf edge, creating a series of folds, Springmount

Beech: *Stigmella hemargyrella*, (June and August to September). *Stigmella* mines are fun to look out for and we must watch for the shape of the mine and the frass (poo). In this one, the frass is coiled in the central part of the mine. The egg at the start of the mine, is positioned away from the midrib of the leaf, which distinguishes it from *S. tityrella*. Larva is white-yellow, garden

Beech: *Stigmella tityrella*, (June to July and August to October). Another *Stigmella* mine, which has no coiled frass. In this case, the egg is on underside of the leaf against the midrib. The larva is yellow, with a dark head, garden

Step into Nature

Blackthorn: *Stigmella plagicolella*, (July and September to October). This is such a distinctive leafmine, starting with a narrow gallery which develops into a blotch, often referred to as the Tadpole mine. Larva is pale yellow with a brown head, garden

Blackthorn: *Phyllonorycter spinicolella*, (July and September to April). *Phyllonorycter* mines are often characterised by creases on the leaf surface. This mine has several creases in lower epidermis which causes the leaf to arch or fold over, garden

Hawthorn and Rowan: *Stigmella oxyacanthella*, (September to October). A great mine, with initially linear, reddish frass, later it is in distinctive arcs, finally dispersed. The larva is bright green, garden. To view these mines clearly, the leaf should be turned up towards the light (still attached to the branch)

Hawthorn: *Phyllonorycter oxyacanthae*, (July and September to October). Another *Phyllonorycter*, it mines the leaf lobe, causing many creases to the lower epidermis. It becomes strongly contracted causing the lobe or leaf-edge to fold over, garden

September - Week 4

Mushrooms come in such vibrant colours and amazing shapes and forms. But the Fly Agaric with its white spotted, shiny scarlet cap is synonymous of childhood fairytales. The spots are actually fleecy scales, that wash off with rainfall over time. The white cylindrical stipe has a bulbous, scale-covered base and a pendulous ring with a scaly edge. They are widespread in acidic woodland with Birch trees. Highly toxic, the generic name comes from the medieval custom of placing parts of the cap in milk to stun flies. They belong to a broad group of *Agaricus* or Wood Mushrooms which include the common favourite field mushroom, *Agaricus campestris*.

Also vibrant, the graceful form of Yellow Stagshorn is found firmly attached to conifer stumps or roots.

Fly Agaric, *Amanita muscaria*, (stipe 23cm, cap 20cm, late summer to late autumn), Lough Boora

Yellow Stagshorn, *Clavaria viscosa*, (8cm high, autumn), Lough Boora

Other varieties found in gardens include the Shaggy Ink Cap or Lawyer's Wig. As this tall Inkcap matures the cap opens out at the base and inner layer of gills begin to deliquesce (auto-digest). The spores ripen first from the base, turning to black fluid which drips to the ground bringing the spores with it. The liquid was and can be used as a black ink and when mixed with cloves and water, fixes the colour. They are commonly found in lawns, or disturbed ground.

The Deceiver is aptly named as it is so variable in appearance, from tiny button forms to conical, flat-topped or upturned caps. Stems also vary and can be straight, twisted or compressed. Peachy-red in colour they pop up singly or in groups around the base of deciduous trees in the garden or woods.

Shaggy Ink Cap, *Coprinus comatus*, (stipe 30cm, cap 15cm, April to November), garden

Deceiver, *Laccaria laccata*, (cap 1.5-6cm, stipe 6-10cm, summer to early winter), garden

September - Week 4

Aphid parasitised by *Dyscritulus* sp., garden

Have you ever noticed on the underside of a leaf a small aphid that appears to be stuck on a circular base? The aphid has been parasitised by an Aphidiinae (Braconid) wasp, *Dyscritulus* sp. The wasp lays its egg within the aphid, and while most species pupate inside the mummified fly, this example pupates beneath in a larval cocoon (August - W3). With fewer insects at this stage of the year, we may take more notice of caddisflies as they rest during the day. *Halesus radiatus*, with distinctive, streaked wing markings, is common close to streams, lakes and ponds. This small, beautifully marked water beetle, *Sphaeridium* sp., is equipped for land and for swimming. It is a dung feeder and females lay their eggs in fresh dung where larvae develop. They pupate in four weeks, with adults emerging from early June.

Halesus radiatus, (17-22mm, August to November), garden

Step into Nature

Sphaeridium sp., (5-7mm, all year), on grass, garden

October

Week 1

Few insects move with such speed as the Humming-bird Hawk-moth, as it flits from one flower to another. While hovering, it uncoils its proboscis (the length of its body) to collect nectar. This is achieved by flapping its wings up to 85 beats per second, allowing it to hover.

Humming-bird Hawk-moth, *Macroglossum stellatarum*, (20-24mm, August to early October), migrant

Note the proboscis extended as it feeds on Aubretia in sunshine, garden

Step into Nature

The Green-brindled Crescent positively sparkles against a dark backdrop of brown, to blend with lichen and moss-covered timbers. Wings are dusted with bright metallic green scales that shimmer in the light. Mary Ward of Ballylin, Offaly, in her captivating book *Sketches with the Microscope*, examined such scales of moths and butterflies in 1857. The microscope shows tufts of stunning green, which over time wear off an older moth. Eggs are laid on Hawthorn, Blackthorn or Crab Apple twigs, and overwinter. Larvae pupate underground in a cocoon.

Green-brindled Crescent, *Allophyes oxyacanthae*, (17-20mm, late September to early November), garden

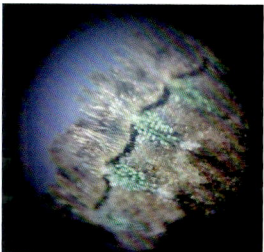

Wing detail of sleeping moth, viewed under microscope, highlighting green scales

October – Week 1, the fruits of Autumn

Hawthorn, Crab Apple, Guelder Rose,
Blackberry, Blackthorn, Bush Vetch
Wild Rose, Ivy, Spindle, garden

Step into Nature

Many of our garden or hedgerow plants are now displaying their colourful fruits, dry or succulent, so designed to attract the chosen bird or insect who will carry out the important task of scattering their seeds. They provide such a vibrant, cheerful display that we have no reason to miss their flowers.

Other unusual sights are some current mushrooms. Earthstars are peculiar onion-like fungi. The outer layer folds back to the ground, revealing the central sac which releases spores, similar to a Puffball (September - W3). The most familiar is the Collared Earthstar which grow in soil rich with humus. A particular favourite are the White and Elfin Saddle. Such a curious structure as they push up through the ground with their furrowed stem of hollow chambers and irregular, distorted head. They are common in both deciduous and conifer woodlands.

Collared Earthstar, *Geastrum triplex*, Birr Demesne

White Saddle, *Helvella crispa*, (15cm), Birr Demesne

Elfin Saddle, *Helvella lacunosa*, (15cm high, summer and autumn), Cranberry Bog

October – Week 1

One of our most endearingly familiar, doggedly persistent and beautiful flowers is the Daisy. As a member of the huge Asteraceae family, it can be used to illustrate the complexity of what we take to be a single flower. In fact, it is made up of around 250 individual florets; female ray florets (white 'petals') with a pink-tipped underside, and hermaphrodite disc florets (yellow centre) which have both male and female parts. The central disc florets are bell-shaped and unfurl up directly towards the sun. They would be damaged by rainfall, so the outer ray florets come to their aid and close over in cloudy weather and at night, protecting the inner florets from rain and dew. Rev. Johns in his illustrated book, *Botanical Rambles*, describes how on a still, wet day, the white ray florets close together allowing water run off the outer surface. But if it is windy, they flex with the wind and the base of the flower takes the brunt of falling rain.

As we know from garden lawns, if the flower is cut, another may be produced from the centre of a basal rosette of thick leaves. The rosette grows tight to the ground, so designed to prevent grazing, but thankfully, equally effective against a lawnmower. The flowers are visited by a variety of insects including flies, bees, beetles and butterflies.

Step into Nature

Daisy, *Bellis perennis*,
(10cm high, flower c.20mm,
February to October), garden

October – Week 2

Week 2

We have been introduced to some garden moths decked out in autumnal colours, but there are also a large range of those with subtle darker shades. A dusty, none too common, Figure of Eight blends perfectly with lichen covered timbers. It overwinters as an egg, laid in small groups on Blackthorn, Hawthorn or Crab Apple. The Lunar Underwing, with noticeably pale veins, overwinters as a hardy larva, low in the grass which feeds by night. The softly coloured Grey Shoulder-knot is one of the few moths that overwinters as an adult. October sun washes the beautiful Red-line Quaker and a pinkish form of Blair's Shoulder-knot.

Figure of Eight, *Diloba caeruleocephala*, (15-19mm, late September to mid November), with pale green 8 on the wing

All found in the garden, many were county firsts, highlighting the importance of recording

Step into Nature

Lunar Underwing, *Anchoscelis lunosa*, (14-17mm, late August to mid October). Habitat - a range of grasslands, parklands and gardens, and sometimes found in the evening resting on fences or on grasses

Grey Shoulder-knot, *Lithophane ornitopus*, (17-19mm, September to November and February to April). Feeds after dark on Ivy and berries, and on Willow catkins in spring, found by day

Red-line Quaker, *Leptologia lota*, (15-18mm, September to November), also feed on Ivy and berries

Blair's Shoulder-knot, *Lithophane leautieri*, (17-20mm, late September to November), overwinters as an egg

October - Week 2

Papillose Bog Moss,
Sphagnum papillosum, (25cm)

Red Bog Moss,
Sphagnum capillifolium, (15cm)

Feathery Bog Moss,
Sphagnum cuspidatum, (15cm)

Lustrous Bog Moss,
Sphagnum subnitens, (20cm)

Broom Fork Moss,
Dicranum scoparium, (10cm)

Pointed Spear-moss,
Calliergonella cuspidata, (12cm)

Mosses, like ferns, reproduce by spores but in capsules; either as sphagnum, ejected forcibly through a small opening, or when a lid drops off, as in most mosses. They have stems but no roots, and leaves generally in spirals. The following mosses were found at Mongan Bog, close to Clonmacnoise.

Papillose Bog Moss is a pale green and yellow with hoods and cortex (outer layer of stem cells). One of a number of Sphagnum, it is common at the base of hummocks in wet acidic habitats, forming tussocks of buff green. Red Bog Moss is a red pink, with no hoods and no cortex. Found on bogs or marshes, it forms a crimson carpet tinged with green under Heather or on hummocks. Feathery Bog Moss has thin leaves and favours acid bogs, forming yellow-green mats. It may also be aquatic, submerged in pools or lakes where it can grow twice as long. Lustrous Bog Moss is a dull pink and can be red or brown with pointed tips to branch and stem leaves, it is common in blanket bogs.

Broom Fork Moss is a fresh lime green, feathery with a mid-rib. This moss is happy in a variety of habitats, from rocks to trees, to stumps and soil. Single fruit stalks are yellow with a red base. Pointed Spear-moss is a common moss on grassland, lawns, wet bogs and even damp areas of sand dunes.

October - Week 2

Campodorus sp., Ichneumon wasp, scouting for sawfly larvae

A bright afternoon brought us to the Cranberry Bog, and on a small Willow sapling, we noticed several *Euura pavida*, sawfly larvae. Scurrying around from leaf to leaf, there was also a female Ichneumon wasp, *Campodorus* sp. A primary parasite, she oviposits into the larvae. Her eggs will develop inside and emerge as larvae. As we have seen before (July - W4 and August - W3), they may, in turn, fall prey to a secondary parasitoid or a hyperparasite, possibly another Ichneumon, *Mesochorus* sp.

Campodorus sp., Ichneumon wasp, parasitizing *Euura pavida*, sawfly, on Willow, Cranberry Bog, Gallen

October shieldbugs   *Step into Nature*

Birch Shieldbug, *Elasmostethus interstinctus*, (8-11.5mm, September to June), garden

Birch Shieldbug nymph, also October

Bronze Shieldbug, *Troilus luridus*, (10-12mm)

Green Shieldbug, *Palomena prasina*, (12-13.5mm)

Hairy Shieldbug, *Dolycoris baccarum*, (11-12mm)

Hawthorn Shieldbug, *Acanthosoma haemorrhoidale*, (13-15mm, September to July), laying eggs on Rowan leaves

Hawthorn Shieldbug nymph, *Acanthosoma haemorrhoidale*

October – Week 3

Penny Bun, *Boletus edulis,* or Birch Bolete, *Leccinum scabrum* — 342 — (stipe 15cm, cap up to 25cm across, summer to late autumn), Clooneen Bog

Step into Nature

Week 3

When correctly identified, a fresh Penny Bun or cep, is greatly sought after for its culinary excellence. Belonging to the *Boletus* genus, the cap underside is made up of tightly packed tubes rather than gills. They have such a statuesque appearance, reminiscent of a noble tree in a miniature wood as they grow in deciduous and coniferous woodland. Oysterlings, on the other hand, have the most magnificent gill construction. Another broad family, they are found in clusters on decaying deciduous timbers. The fan-shape cap with wavy margins hides the stunning architecture beneath.

Oysterling, *Panellus* sp., (c.4cm), Kinnitty Demesne

October - Week 3

Leucoptera laburnella, (June, July and September to October), circular or oval blotch with spiral frass, on Laburnum, Birr Demesne

Leucoptera lotella, (June to July and August to mid-October), also circular blotch with spiral frass, on Bird's-foot-trefoil, Cranberry Bog, Gallen and Ballylin Bog

Acrolepia autumnitella, (June and September or October), a large translucent mine, larva ready to pop out, on Bittersweet, *Solanum dulcamara*, Birr Demesne

Phyllonorycter nigrescentella, (May to June and September to October), underside of leaflet turns down, multiple creases, on Bush Vetch, garden

Step into Nature

There is still time to find interesting bugs. Damselbugs are widespread in damp grassland and voracious predators. A slight movement in long grass causes them to rush out in the hope of prospective prey. The Broad Damselbug has a wide body with dense golden hair, and forewings half the length of its abdomen. Eggs are laid in grass stems in summer, hatching next spring with adults developed from June. A similar Marsh Damselbug has slightly shorter wings.

Broad Damselbug, *Nabis flavomarginatus*, (7-9mm, June to October), Birr Demesne

Marsh Damselbug, *Nabis limbatus*, (7-9mm, June to October), Killaun Bog

The most unusual looking Thread-legged Bug (assassin bug) hunts for insects, spiders and aphids, on trees and associated vegetation in shady places. Their shorter front legs fold up, perhaps ready to catch their prey. Black eggs are laid in May, with larvae found along tree branches, adults emerge in July and overwinter in conifers or dead leaves on twigs.

Thread-legged Bug, *Empicoris vagabundus*, (6-7mm, all year), Birr Demesne

October – Week 3

Speckled Wood, *Pararge aegeria*,
(47mm male, 50mm female wingspan,
April to October), Birr Demesne

Step into Nature

The Speckled Wood is such a reliable butterfly of gardens, woods and hedgerow. One of the first to appear in April and we can hope to see them right into October. This beautiful Speckled Wood was basking, low down, in afternoon sunshine. The sun highlighted subtle tones of blue and purple, so often missed in the height of summer, against a deep brown and cream-spotted base. They can have two broods, one overwinters as a pupa and are first to emerge in April. A second brood overwinters as a caterpillar emerging in July and sometimes they produce a third brood which appears in October, as this one. When we see two spiralling around each other, it is usually two males disputing territory. Females lay eggs in wild grass which hatch within two weeks. Initially pale green with long hairs, the caterpillars develop dorsal stripes and a white 'tail', (as this one found in April). They transform to a green pupa which lasts at least two weeks depending on the season.

Speckled Wood, *Pararge aegeria*, bright green larva, on grass, Killaun Bog, April

The same larva as found, blending into the grass, can you find it?

October – Week 4

Week 4

For this week, we are going to concentrate on fungi and also take a look at slime moulds. There are so many varieties to choose from, both in the garden and in woodlands. Dark Honey Fungus grows in patches on the lawn and is widespread and common. It grows on both living and dead wood, either deciduous or coniferous. An unusual garden find is the Field Bird's Nest, aptly named as they look like a tiny nest equipped with 8-10 flat, shining eggs. These are spore packets within the cup-shaped fruit body and all it takes to release the spores is a splash of rain. Another type of Puffball (Common Puffball, September - W3), but much larger, is the Pestle Puffball. Initially cream with spiny warts, it later loses the warts and turns brown. The head splits and spores are released.

Dark Honey Fungus, *Armillaria ostoyae*, (stipe 15cm, cap 15cm), garden

Field Bird's Nest, *Cyathus olla*, (1-1.5cm across, late summer to autumn), in woodchip, garden

Step into Nature

Pestle Puffball, *Lycoperdon excipuliforme*, (up to 20cm tall, late summer and autumn), garden

October – Week 4

Blue Roundhead, *Stropharia caerulea*, no scent, (stipe 10cm, cap 3-8cm), garden

Blue Roundhead, *Stropharia caerulea*, spore print

1. Clouded Funnel, *Clitocybe nebularis*,
2. Sulphur Tuft, *Hypholoma fasciculare*,
3. Tawny Funnel, *Lepista flaccida*,
4. The Deceiver, *Laccaria laccata*,
5. Parrot Waxcap, *Hygrophorus psittacinus*, cap of each (left) and spore print

Step into Nature

Blue Roundhead is a super garden find which reappears each year during the last weeks of October. An incredible blue-green, slimy with white scales, it grows in mulched flowerbeds. The stipe has a distinct ring, smooth above and covered in white scales below. Under the cap, the gills are buff-coloured and as the mushroom ages, the gills turn a darker brown. Each gill can hold thousands of spores, and we can see these by making a spore print.

Spore prints are great fun and give a perfect imprint of each mushroom. We cut off the stipe (or cut a hole in the paper) and place the cap, gills down on a sheet of white paper. A drop of water on top of the cap helps to speed up the release of spores, and a glass over the cap keeps them contained. Over the next two to 24 hours, the spores fall onto the paper leaving a unique pattern, either airy or relatively solid.

Each species has different colours which make fascinating prints. Blue Roundhead leaves a dense brown print while others are pink, rust, yellow or even black. The Roger Phillips book, *Mushrooms*, notes the spore print colour of each variety. Coloured card can provide a vibrant contrast and the print can be fixed permanently with a light mist of hair spray (from a distance) - a perfect greeting card.

October – Week 4

Slime moulds are often confused with fungi and, in fact, used to be thought of as a strange form of fungi. But they are entirely different and hugely varied, lasting just a few days at peak condition. They can change significantly as they develop and age, which makes them even more interesting and also difficult to identify. When we first encountered the tiny form of *Trichia varia*, along a damp, moss-covered fallen timber, it was bright white, similar to snails eggs. Four days later, they turned orange, ready to fruit, releasing spores. An even more fascinating example is *Ceratiomyxa fruticulosa*. Ghostly translucent tubes grow in clusters across dead wood (even a garden post), like spreading ice. They are truly magical.

Trichia varia, (head less than 1mm, stalk up to 0.5mm), Kinnitty Demesne

The same *Trichia varia*, four days later

Step into Nature

Ceratiomyxa fruticulosa,
(June to October),
Kinnitty Demesne and garden

November

Week 1

While October fades away into November, we can look forward to vivid displays of colour, as leaves turn from vibrant green to yellow and fiery orange to red. During an early morning visit to Killaun, we were fortunate to encounter a silent herd of elegant Fallow Deer. Known as even-toed ungulates (with divided hoofs), they tread lightly through the bog, feeding on young shoots of trees and shrubs. When numbers are high, they can cause damage to forestry by browsing on saplings, bark and shoots. Introduced to Ireland by the Normans for food and hunting, they are similar in size to sika deer and smaller than red deer.

Fallow Deer, *Dama dama*, Killaun Bog

Single fawns, born in the summer, vary in colour from a red coat with white-yellow spots to almost white or black, darkening in winter.

Looking now at Oak trees, we are sure to find *Andricus kollari* the gall wasp. It causes either singular or clusters of the smooth marble gall. Modified green buds with a central chamber containing a single larva, turn brown and woody. Adult wasps leave the gall in September via a neat circular hole. Additional holes indicate the presence of parasitoids. *Andricus curvator* cause spherical galls within young twigs or in groups near the leaf base, causing distortion as they grow. The impressive *Andricus quercuscalicis* have a large chamber enclosed in a ridged gall with a further inner chamber and a single larva.

Andricus kollari, (20mm), gall wasp on Oak buds, garden

Andricus curvator, (2-3mm), gall wasp on Oak leaves or twigs

Andricus quercuscalicis, (20mm), a Knopper gall wasp on acorns

November - Week 1

November is a good time to see Snipe, a resident, bolstered by migrant birds from northern climes. It is a relatively common wader with an extremely long, straight bill. Ireland is fortunate to have summer visitors from west Europe and west Africa. In the winter, we have visitors from the Faroe Islands, Iceland and northern Scotland.

Snipe nest on grassy tussocks in wet areas, marshes, bogs, river and the lakeshore. It is in these locations that they are so well-camouflaged, with dark brown and black stripes, buff and cream markings that blend into the surrounding reeds or marshy vegetation. They feed mostly on vegetable matter, worms or larvae unearthed by rapid probing with their long bill (known as the sewing machine movement).

During the winter, their distribution is classed as highly dispersed by BirdWatch Ireland. They forage across a variety of wetland habitats. Damp areas and the fringes of lowland lakes are favoured, we had one such visitor to the garden in early November.

Snipe, *Gallinago gallinago*, (length 25-28cm)

Step into Nature

If the weather is mild, tall and stately Purple Toadflax continue to flower into November. It grows by roadsides, on walls and self-seeds in gardens, where dense spikes of purple flowers are a wonderful addition. Each flower is similar in structure to the Snapdragon, though considerably smaller.

Purple Toadflax, *Linaria purpurea*,
(30-80cm high, June to November),
garden, photo taken in June

November – Week 1

Swarms of bright, juicy Lemon Disco spring up on dead, deciduous wood. Smooth and slightly rubbery, widespread and said to be common, it is always a cheerful treat enlivening the immediate area. Purple Jellydisc is spherical, cushion-shaped with raised edges. Smooth, soft and unsurprisingly gelatinous, it is a beautiful pinky-purple with darker margins. They are typically found in dense clusters on rotting deciduous trees, especially Beech. Coral Spot is an easy one to find on dead branches around the garden or in woods. There are two types, cinnabar-red flasks and the more familiar soft, coral-pink raised spheres.

Coral fungi are so elegant and give the appearance of an underwater seascape. This upright, ochre coral has long, smooth, dense branches that repeatedly sub-divide as it grows on decaying timber or mulch.

Lemon Disco, *Bisporella citrina*, (3mm across, saucer or flatter disc), and Purple Jellydisc, *Ascocoryne sarcoides* or *A. cylichnium*, separated by microscope (10mm across), Kinnitty Demesne. Coral Spot, *Nectria cinnabarina*, (4mm, all year), garden

Step into Nature

Upright Coral, *Ramaria stricta* or *Phaeoclavulina* sp., (4-10cm high, late summer to winter), Kinnitty

November - Week 2

Ichneumon sarcitorius, male, (10-12mm), garden hedgerow

Step into Nature

Week 2

This magnificent *Ichneumon sarcitorius* was a treat to find on a chilly November morning and extremely conspicuous due to its size and striking colouring. Females have brown legs and predominantly brown stripes to the abdomen, males have yellow legs and a bright mix of white, yellow and orange stripes. They have the characteristic narrow waist of wasps, but exceptionally long antennae, made up of 16 or more segments. While aware of my scrutiny, this male was perfectly happy to stay still, posing with noble stature on a Nettle leaf.

The rather rare, plush Sprawler was an unexpected and very welcome visitor to the garden. It overwinters as an egg, and yellow-green caterpillars with three white lines feed on broadleaved foliage May to June.

The Sprawler, *Asteroscopus sphinx*, (17-22mm, October to December) garden

November - Week 2

Feathered Thorn moths are a common species of rich autumnal colouring, that rest with their wings flat, unlike the Early Thorn. Males are distinguished by their feathered antennae, and may be found during the day resting on shrubs or at the base of tree trunks. They overwinter as a batch of olive-green eggs with a delicate speckled ring at one end. Eggs are laid on their larval foodplant of deciduous, preferably broadleaved trees in woods, hedgerows and gardens.

Feathered Thorn, *Colotois pennaria*, (19-23mm, September to December), garden

Feathered Thorn egg detail, batch on branch, pupa and larva

Step into Nature

Purple-grey-brown larvae appear April to June, with a row of dull yellow spots down each side, diamonds shapes along their back and two red rear projections. They feed at night and pupate just below ground.

On a far bigger scale, we must consider one of our larger resident birds. The statuesque Grey Heron is most commonly seen standing still by a pond or river, sheltered by low-hanging branches and surrounding vegetation. It waits patiently, immobile and disregarded by the unsuspecting fish below. Then with a dart, the graceful neck dips down and the long yellow bill snatches its prey. They are formidable hunters with a long memory, and once they discover a source of food, it will remain on their flightpath. They scan while flying overhead, with their neck retracted and long, green legs trailing behind.

Grey Heron, *Ardea cinerea*, (90-100cm), feed on fish, insects, frogs and small mammals

November - Week 2

More fungi of superb colours grace our grasslands, forest floor or garden, making up for any perceived paucity of winter months. A firm favourite is the delightful, viscid Parrot Waxcap, which, each year seems smaller than previously remembered. They grow at the base of an old garden Ash tree, barely peeping over the surrounding moss. Waxcaps or *Hygrocybe* are an incredibly varied and colourful genus with some species of vibrant orange, red and yellow. Fibrecaps or *Inocybe* are generally small with more subtle colours. White Fibrecaps have a conical cap, smooth and silky. The lilac form, found at Kinnitty, is particularly beautiful. Wood Blewit are widespread and common, growing in clusters on woodchip mulch.

The enchanting Amethyst Deceiver, of deciduous woodland, has a felty cap of the most surprising violet or amethyst, which fades over time to cream or tan.

Parrot Waxcap, *Hygrophorus psittacinus*, (stipe up to 7cm, cap 4cm), garden
Lilac Fibrecap, *Inocybe geophylla var. lilacina*, (stipe 6cm, cap 4cm), Kinnitty Demesne
Wood Blewit, *Lepista nuda*, (stipe 10cm, cap 15cm), garden

Step into Nature

365

Amethyst Deceiver, *Laccaria amethystina*, (stipe 10cm, cap 5cm), in the embrace of a delicate web, Kinnitty Demesne

November - Week 3

Ectoedemia subbimaculella, (October to November), on Oak, a small gallery mine leads to blotch, distinguished by slit on underside to allow frass be ejected, garden

Stigmella luteella, (August to November), on Birch, first contorted then scalloped, frass free edges. Larva yellow, with green gut-line, garden

Stigmella betulicola, (July, September to November), on Birch, several mines with parallel edges (not scalloped). Larva yellow, gut-line green, small brown spot, Finnamore Lakes

Week 3   Searching for leafmines can seem futile as leaves continue to fall, but sometimes we should look further at those fallen leaves. Some leafminers can induce 'green-islands' on the otherwise yellow leaves, such as *Ectoedemia argyropeza* on Aspen. While the rest of the leaf is senescing (loosing chlorophyll) as normal, the green patches around the leafminer continue to photosynthesise. *Ectoedemia* work symbiotically with bacteria to artificially extend their season on the fallen leaves. The mine, initially in the petiole, moves up and into the green areas when the leaf falls. Larvae overwinter in cocoons and pupate the following spring.

Step into Nature

Ectoedemia argyropeza, (August to November), on Aspen, at petiole, (leaf stem)

Later forms noticeable 'green island' on fallen leaves, Kingstown, Birr

November - Week 3

We first encountered a magnificent Buzzard, resting on a tree in the garden, as we came up the driveway. Smaller than an eagle but significantly larger than a Sparrowhawk or Kestrel, the broad-winged, short-necked raptor gripped a branch with strong yellow talons.

Since then, Buzzards have bred and nested nearby with broods of two to four young each year (June to August). So dramatic in flight, they soar overhead at spectacular heights with adults guiding their young. A distinctive high-pitched call signals their presence and it is mesmerising to watch them circling over the garden with dark plumage and a dark trailing edge with contrasting pale parts to underwing.

Their prey range from rabbits to young birds, frogs to insects. Despite practical extinction, numbers have dramatically increased in recent years and it is comforting to know that they can live into their twenties.

Buzzard, *Buteo buteo*, (wingspan 115-130cm)

Step into Nature

Slime mould, *Metatrichia floriformis*, like tiny black lollipops, on decaying timber at Charleville Demesne

Crystal Brain, *Myxarium nucleatum*, garden

Beech Jellydisc, *Neobulgaria pura*, Wrinkled Club, *Clavulina rugosa*, Kinnitty Demesne

November - Week 4

Week 4

Resting on a leaf by the river in Kinnitty is an insect that we have not yet looked at. Stoneflies belong to the order Plecoptera and are more than 250 million years old. There are 3,500 species worldwide, 570 in Europe, and 20 in Ireland - which belong to seven families. They are exclusively found in freshwater habitats; streams, rivers and lakes, particularly upland, fast flowing with stoney base or banks. They also favour canals, turlough, ponds and reservoirs. Adults are always found on land and larva in water, they are Hemimetobolous, having no pupal stage.

The egg stage may last from a few hours to a year, and are unique to each species. They rely on 10° or lower water temperatures, leading to issues with climate change.

Leuctra fusca, adult, (6-9mm, June to December, nymph 7-10mm, April to November), Kinnitty

Isoperla grammatica, nymph, (9-15mm, all year, adult 8-14mm, April to August), Camcor River, Birr

The nymph stage which may be three to four months or three to four years, is similar to the adult without wings and has two tails (while the mayfly larva has three tails, May - W3). In general, April to May is the best time to see them in water. After shedding their skin they are pale yellow, soft and vulnerable until they harden. Reaching maturity, their wings pads begin to develop. Larvae crawl out of the water onto bank or rock (may be a considerable distance), usually at night. The exoskeleton splits, and the soft adult pulls itself out, pumping its folded wings. There can be hundreds of exuvia on rocks or walls and the cuticle may be used to identify the adult. Adults live from a few days (no mouthparts, just to reproduce) to one month (have mouthparts, and feed on pollen). Females can be three times the size of male.

Stoneflies have a unique system of drumming to find a mate and each species has its own patterned beat. Males drum their rear and the female answers from afar, with constant drumming they close the distance until they find each other. Once the female mates, she does not drum again.

After mating, males guard females to ensure another male does not try to mate with her. Some males have no wings and those who do are not fond of flying.

November - Week 4

They often rest on leaves in sunshine or fence posts. Females have to fly to water and lay eggs, dipping their abdomen into the water as eggs drop in.

Stoneflies play an important role in nutrient cycling, by breaking down leaves in the water. Larvae provide food for fish and birds, (e.g. Salmon and Dippers), spiders and other terrestrial invertebrates (ants). They are pollution sensitive, so indicators of good quality water and survive in acidic water, unlike mayflies.

Protonemura meyeri, garden in April, adult (5-9mm, April to November, nymph 7-10mm, June to April)

For more detailed information refer to, *The Stonefly (Plecoptera) of Ireland*, by Feeley, Baars and Kelly-Quinn

Step into Nature

While checking by the pond, we found the Brick moth struggling on the surface. This was the first time we encountered the buff-coloured moth with rusty tinges. Overwintering yellow eggs, like minute sea urchins, are laid in batches on buds and, after a week, turn purple with a pearly sheen. The yellow-brown caterpillar (April to June) has a red head with a mix of stripes and V-shaped markings. It feeds on Ash flowers first, then moves to seeds and leaves. It makes an underground cocoon pupating six weeks later. Adults feed on Ivy flowers, berries and Heather. Another linked first, found indoors, was the micro-moth *Mompha subbistrigella*, which hibernates as an adult. They live by ponds or ditches and their red larvae live within Willowherb seedpods.

Brick, *Sunira circellaris*, (14-19mm, late August to December), garden, next to find - the caterpillar!

Mompha subbistrigella, (3.5-5.5mm, all year), indoors, we must look out for the red larvae

November – Week 4

A cold November weekend is the prefect time to seek out wintering birds, at Lough Boora or Lusmagh bird hide for example. Shakily taken through a spotting scope with a smartphone, we got record photographs of the some of the many visiting species. Lapwings are quirky resident waders with a jaunty, flicked crest. In flight, their conspicuous white breast and white underwing with black edges are very distinctive. They favour open habitats, grazed grassland or arable farmland and sweep over their territory, uttering a unique call in spring. We spotted two at a neighbouring farm. Large flocks are common in winter at suitable feeding sites. They feed on invertebrates, often by night, and nest precariously on the ground in bare, cultivated fields.

Golden Plovers are plump, gregarious, beautifully marked waders, who gather in large flocks. They have a more varied diet of invertebrates, grasses, seeds and berries, and breed in Heather bogs or grassland.

Viewed through a long scope, Lapwing, *Vanellus vanellus*, (28-30cm), and Golden Plover, *Pluvialis apricaria*, (26-28cm), Lough Boora

Mallard, *Anas platyrhynchos*, (50-60cm), Banagher

Step into Nature

Mallards are perhaps our most familiar resident and migrant duck - both the colourful male with gleaming green head and yellow bill, and brown-flecked female. They feed on seeds, grains and crustaceans, and nest on the ground in vegetation. Our most common wetland bird, the black Moorhen with a bright red and yellow bill, swims pushing its head forward and back, while flicking its tail. They are very timid and dash for cover in emergent vegetation when startled. We had a visiting pair in the garden with chicks over a few years, where we first noticed their yellow legs and surprisingly oversized, long yellow toes. They feed on fruit and seeds, insects, tadpoles and fish.

Stonechats, small songbirds, are vivacious little birds. Males have a black head with white sides and a orange breast, females are brown streaked. Favouring scrubby areas with Gorse, they perch on trees or posts, uttering their brusque call.

Moorhen, *Gallinula chloropus*, (32-35cm)

Stonechat, *Saxicola torquate*, (12-13cm), male (left) and female (right), Lough Boora

December

Week 1   A firm favourite popping up under Beech trees, close to the base of the trunk, are these gorgeous Jellybaby fungi. They vary in colour from ochre to orange to olive green, growing gregariously through moss and fallen leaves.

Jellybaby, *Leotia lubrica*, (up to 6cm tall, August to December), garden

Step into Nature

Next we check on Holly as there are a number of things to look out for, including its only leafminer. *Phytomyza ilicis* is a fly, belonging to the same family as bluebottles and houseflies, but only 2.5mm. Eggs are laid late spring or early summer on the underside of the midrib (red spot). Hatched larva work slowly though the midrib (5mm in 3 months), moulting in December. It then turns and works its way through the leaf blade, forming a blister. Young larva feed through to the following spring and pupate in April or May.

Adult flies extract themselves from the tightly packed pupa by distending a bladder-like protrusion on their head, which splits the outer shell. They emerge from the leaf through a hole pre-cut by the larva, in May or June. Astounding forethought.

Holly Leaf Gall Fly, *Phytomyza ilicis*, (spring, summer and winter), garden

Holly Speckle, *Trochila ilicina*, a dark disco fungus (3mm), on upper side of fallen Holly leaves, Killaun Bog

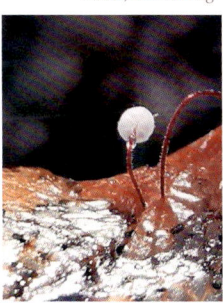

Holly Parachute, *Marasmius hudsonii*, (5mm), often solitary on moist fallen leaves, Killaun

December - Week 1

Great Tit, *Parus major,* (14-15cm)

Due to much colder temperatures, our smaller birds really appreciate additional seeds or nuts, and we can enjoy their visits to the bird feeder. The bolder Great Tit is often first up, with white cheeks, black head and throat, and black central band to the breast flanked by cheery yellow. Considerably smaller, the dotey Blue Tit has a bright blue head, a white face with a narrow navy eye-band and a full yellow breast. It nests in tree-holes and both the Great Tit and Blue Tit are happy to make use of bird boxes in the garden. Interestingly, Blue Tits feed on the Holly leafminer, making a small triangular flap in the epidermis to extract the larva. Similarly petite, but slighter, the Coal Tit has white cheeks to an otherwise black head and a small white nape patch. It has a pinky-buff breast with dark grey wings. It is shown here trying to visit the bird feeder, which is not easy when rowdy Greenfinches are around.

Coal Tit, *Periparus ater* (11-12cm), and Greenfinch, *Carduelis chloris,* (14-15cm)

Blue Tit, *Cyanistes caeruleus,* (11-12cm)

Step into Nature

Predominantly yellow-green with a vivid yellow streak to the wing and tail, they have a dark patch around the eye which seems to lend them a permanently grouchy air. They certainly are the most belligerent visitors to the feeder.

One which can easily be overlooked is the Dunnock. Similar to the Sparrow, (January - W1), but more muted grey-brown, it is cautious, preferring to feed close to ground cover, thus more difficult to observe. There can be no such difficulties with the brightly-clad sociable Goldfinch with their bright red face encircled with white. A firm garden favourite, the diminutive Wren with its upturned tail, flits around low vegetation in search of insects. Males make several domed nests and bring prospective mates to visit each before they make the final decision.

Dunnock, *Prunella modularis*, (13-14cm)

Goldfinch, *Carduelis carduelis*, (13-14cm) and Wren, *Troglodytes troglodytes*, (9-10cm)

December – Week 2

Week 2

We have over 20 species of ants, from the Formicidae family, in Ireland. Social insects, hundreds or thousands of wingless workers form colonies that last several years. Those seen outside the nest are usually workers collecting food to rear the grubs. Only males and queens (reproductive females) have wings, large clear or tinted brown. New males and the larger new queens (up to twice the size of workers) pour from the nest in great swarms on warm, humid days in summer. They mate in flight and males die soon after. Egg-laying Queens shed their wings after mating and must either start a new colony or hope for acceptance in an existing anthill, with workers to care for her eggs and larvae. Most ants are omnivores, feeding on nectar, aphid honeydew (for protection offered), seeds and living or dead insects. Nests persist throughout the year, but in winter, the ants keep to their nests, in a dormant state.

Species include *Formica* sp., *Myrmica* sp. and *Lasius* sp., (3.5-9mm, yellow-and-black)

Formica sp. (from 3.5-11mm, black-and-red), and *Myrmica* sp. (3.5-5.5mm, red)

Step into Nature

Irish ladybirds are a hardy group, and adults overwinter, often unprotected in the open. Our last two for the year include another yellow species, the 22-spot Ladybird. Smaller than the 14-spot (March - W1), they are bright yellow with 22 (or 20) rounded black spots. They live in grasslands, meadows and gardens feeding on aphids and, surprisingly, powdery mildews. 10-spot Ladybirds are normally red with ten black spots and a white scutellum and head. But there are many variations making them most interesting. Found in deciduous trees, hedgerows and gardens, they overwinter in leaf litter to emerge in spring.

22-spot Ladybird (yellow), *Psyllobora vigintiduopunctata*, (3-4mm), top left

10-spot Ladybird, *Adalia decempunctata*, (3.5-4.5), a selection of typical variations

December - Week 2

Within the most common of insects there is always beauty. The Earwig, (Dermaptera), is one which we seldom consider. Yet they are a fascinating group and display far more maternal care than most of our insects. While there are over 2,000 species of earwig worldwide there are just three in Ireland; the Common Earwig (10-15mm), Lesne's Earwig (6-7mm), with no hind wings and just a few confirmed records, and the diminutive Lesser Earwig (4-6mm), found sporadically in the south and east coast.

The hindwings of the Common Earwig extend slightly beyond the elytra, males have curved forceps to the rear which are longer than the straighter female version. These are used for defence against predatory insects. Adults hibernate over the winter and emerge in spring and from then we can find a nest or brood chamber, with eggs (batches of 30-50) or pale cream nymphs. Eggs and early nymphs are guarded by the mother until she dies (then they eat her). They moult five times until fully mature by July. Shy, nocturnal omnivores, they feed on insect larvae, soft plants and honeydew, hiding under loose bark during the day.

Common Earwig, female, *Forficula bipunctata*, (10-15mm)

Step into Nature

They rarely fly but have clear ribbed wings, tucked under the elytra in a highly sophisticated, fan-like, triple-fold manner. Mary Ward illustrated this phenomenal complexity in *Sketches with the Microscope*, and highlighted the beauty of their delicate rainbow-tinted wings. They fold from the outer edge inwards, then turn up in half, that section flips around to the back, then folds up against the top edge, tucking in under the elytra.

Sometimes we come across a wary Pill Millipede in leaves beneath beech trees. It has a beautiful shiny black finish with pale defined edges to its plates, each of which have two pairs of legs (17-19 pairs in total). When disturbed, it rolls up tightly into a ball.

They play an important role, breaking down dead vegetation.

Detail of Common Earwig wings

Pill Millepede, *Glomeris marginata*, (c.20mm long, 8mm wide), Knockbarron and Charleville Demesne

December – Week 3

A group of ground bugs, *Chilacis typhae*, (4-5mm, all year), on Bulrush, Cranberry Bog, Gallen

Step into Nature

Week 3 For the last weeks of the year, what can we look out for? By water on a sunny day, as the breeze riffles through Bulrushes, the ground bug *Chilacis typhae* will usually be spotted tucking into seed heads. They are easily found as the sun glints off their folded wings. Adults overwinter in the heads, the following spring they lay single eggs on the seeds. Red-brown larvae develop over six weeks and the new adults fly off in search of fresh seed heads. Formerly restricted to Dublin, now found across Ireland, in 2021, this was a first record for Offaly.

Some interesting garden spiders live through the winter including the well-marked *Neriene montana*, found on low vegetation in shady places. It overwinters in leaf litter, Ivy or on tree trunks. *Pachygnatha clercki* is a medium sized, timid spider lives in damper habitats. *Zygiella x-notata*, with a silvery lustre, is found at window and door frames.

Neriene montana, (4-8mm) and *Pachygnatha clercki*, (5-6mm) on Nettle, both in the garden

Zygiella x-notata, (3-7mm), garden, it builds a web with a quadrant missing

December - Week 3

Great Pond Snail, *Lymnaea (Lymnaea) stagnalis*, (50-60mm tall), garden pond

Brown Lipped Snail, *Cepaea nemoralis*, (18-21mm across)

Strawberry Snail, *Trochulus* sp., (10mm across), Clonmacnoise

The Great Pond Snail is a familiar sight in the garden pond as it ambles along underwater in search of food; plants, a dead frog or fish. Their egg capsule (2-6mm), contain up to 300 eggs. Their large shell or house can grow as the snail does. It is continuously added to by the snail's mantle or covering, turning the mouth of the shell from left to right, twisting the spiral tube around a central column.

Glass Snail, *Oxychilus* sp., (6-15mm across), Cappaghmore

Amber Snail, *Succinea* sp., (15-22mm tall), and Garden Snail, *Helix aspersa*, (25-40mm wide), garden

Step into Nature

It can be difficult to imagine that snails have a stomach, heart, liver, muscles and nerves. They have an unusual tongue (radula), with a staggering 110 rows of hook-like teeth with 109 hooks per row. They move some of these backwards and forwards over food to tear it up. When one row becomes blunt they use the sharper row behind. Snails can be difficult to identify, so possible varieties are shown here, highlighting size and colour. There are, of course, predators, including the dramatic-looking Sieve-winged Snail Killer, found earlier in the year. A relatively large fly with broad wings found in damp grassland and hedgerows. While adults feed on nectar, larvae feed on snails.

Sieve-winged Snail Killer, *Coremacera marginata*, (7-10mm long, May to September), Burren

December - Week 4

Week 4

The crow family is full of dark boisterous and highly intelligent birds. Largest and seriously imposing is the Raven, bigger even than our Buzzard, (November - W3). It is all-black with a thick dark bill, and the only crow to have a wedge-shaped tail in flight. A shy species, it pairs for life and breeds in uninhabited woods (or a cliff edge) from early spring. They make an astounding variety of calls.

Rooks have glossy plumage with a blue-purple sheen, a distinctive peaked crown. Adults have a bald patch at the base of their bill, making their bill look significantly longer. They have loose feathers on their belly around their legs, giving the appearance of 'shaggy trousers'.

Raven, *Corvus corax*, (length 55-65cm, common all year, nesting January to February), feed on carrion, eggs, insects

Rook, *Corvus frugilegus*, (46-47cm, very common all year), feed on root crops, berries, insects, slugs, worms

Step into Nature

Familiar nests or rookeries are visible in treetops and their boisterous call is so evocative, when roosting at dusk. These roosts can contain thousands of individual Rooks during the winter months, resulting in a formidable spectacle and a cacophony of sound.

They forage alongside Jackdaws, who are noticeably smaller. Sooty grey in appearance with pale ash-grey neck and back of the head, Jackdaws have distinctive ice blue eyes.

The Hooded Crow occurs throughout Ireland. Similar in size to the Rook, with two-tone colouring of black and pale grey, large roosting flocks visit parks and gardens in the winter. They nest in trees and old buildings. Other members of the crow family include the Magpie, Jay and Chough (confined to coastal areas).

Jackdaw, *Corvus monedula*, (32-35cm, very common all year), feed on insects, slugs, berries, fruit, occasionally eggs

Hooded Crow, *Corvus cornix*, (45-50cm, very common all year), feed on carrion, eggs and young chicks, insects

Only a few resilient moths fly at this late stage in the year. The Winter Moth often rests on a wall close to a light, this will be the male, as females cannot fly. Both may be found on tree trunks after dark. Eggs overwinter on a twig or near a leaf bud. Pale green larvae with delicate cream and dark green stripes feed by day between spun leaves, on a variety of broadleaf woody plants, and pupate in the ground.

One evening only, we were fortunate enough to meet four resplendent December Moths (who came to the light of a moth trap). Magnificently dressed for cold, frosty nights (unlike the more delicate Winter Moth), they are found in woodland, hedgerows and gardens across most of Ireland. Eggs also overwinter, with larvae from April to June feeding at night on broadleaved trees, and resting by day on the bark. We must look out for the larvae in the garden next year!

Winter Moth, *Operophtera brumata*, larva (April to June), and adult (13-16mm, October to January), garden

December Moth, *Poecilocampa populi*, (15-22mm, October to January), garden

Step into Nature

The glorious December Moth,
Poecilocampa populi

Acknowledgements

To my husband and children who bore, with resigned patience, unusual walks, a living room full of books and the scratching of drawings during evening TV, and Gráinne for extra assistance with the cover. To my Mum, sisters and brother who showed such interest (real or feigned) about each new discovery.

The Heritage Officer, Amanda Pedlow first lured me into recording with the bribe of a free NBDC Swatch for 25 records, 'each 25'?! To John Feehan for the thoughtful Foreword, and for his astounding capacity to share his knowledge via books, YouTube Wildflower videos or a highly sought after place at his summer school. He has forever opened a window onto the astounding complexities and beauty of nature around us.

Huge thanks to the dedicated, steadfast experts on the wonderful Facebook pages, MothsIreland, British and Irish Sawflies and Insects and Invertebrates of Ireland. They start each year with the same commitment, providing responses to queries including my very first query about a shieldbug on the clothes line in 2018.

To Jamie O'Neill who read a very early draft, with his all-embracing wisdom and his quiet, easy ability to casually pop in nuggets of information during wonderful, nature-filled walks; as we spend fascinating hours peering at leaves, covering no ground whatsoever.

Step into Nature

Sincere thanks to those who agreed to read through drafts relating to their area of expertise: Brian Nelson, Dave Allen, James O'Connor, John Feehan, Michael O'Donnell, Myles Nolan, Owen Beckett, Ricky Whelan, and a special thanks to Stuart Dunlop and Amanda Pedlow who undertook the entire draft. Their kindness, knowledgeable suggestions and willingness to assist is so appreciated. To Lisa Jewell, for final proofing, any remaining errors are entirely my own, and to Mercier Press for their professionalism and enthusiasm.

Coat Tit, *Periparus ater*

Further reading

January

Week 1: Sterry, P., *Collins Complete Guide to British Mushrooms and Toadstools* (Collins, 2009), p220, 328, 290,
Phillips, R., *Mushrooms* (Macmillan, 2006), p375
Sheldrake, M., *Entangled Life: How Fungi Make Our Worlds, Change Our Minds and Shape Our Futures* (2021)

Week 2: Phillips, R.., *Grasses. Ferns, Mosses & Lichens of Great Britain and Ireland* (Pan Books, 1980), pp98-105
Devlin, Z., *The Wildflowers of Ireland, A Field Guide* (Gill Books, 2021), p34
Pratt, A., *Haunts of the Wild Flowers* (London, 1892), p100, 104, 94
Pratt, A., *Wild Flowers* (London, 1893), vol. ii, p16

Week 4: Devlin, Z., *Wildflowers of Ireland* (2021), p124, 145
Step, E., *Nature Rambles: Winter to Spring* (Warne & Co., 1933), p20
Pratt, A., *Wild Flowers* (1893), vol. i, p59
Sterry, P., *Collins Complete Guide to British Birds* (Collins, 2004), p226
Dempsey, E., *The Complete Guide to Ireland's Birds* (GM, 2002), p202

February

Week 1: www.birdwatchireland.ie
NBDC-Sheildbug-Poster-2019-WEB.pdf (biodiversityireland.ie)
Shieldbugs - illustrated life stages (britishbugs.org.uk)
Gorse Weevil | NatureSpot

Week 2: Chinery, M., *Garden Wildlife of Britain and Europe* (Collins, 1997), pp246-250
Phillips, R., *Grasses. Ferns, Mosses & Lichens* (1980), pp181-3
Step, E., *Nature Rambles: Winter to Spring* (1933), p61
Brimble, L. J. F., *The Floral Year* (Macmillan & Co. Ltd, 1949), p76, pp87-88

Week 3: Irish Peatlands Conservation, Frog factsheet.
Step, E., *Animal Life of the British Isles*, (Warne, 1942), pp157-162
Step, E., *Insect Artisans and their Work*, (London, n.d.), pp197-200
Olsen, Sunesen & Pedersen, *Small Freshwater Creatures*, (Oxford, 2001), pp11-33

Week 4: Feehan, J., *The Wildflowers of Offaly* (OCC, 2009), p168, 170, 155, 221-2
Pratt, A., *Wild Flowers* (1893), vol. i, p25 39

March

Week 1: www.birdwatchireland.ie
Dempsey, E., *Guide to Ireland's Birds* (2002), p210
Sterry, P., *Guide to British Birds* (2004), p232
Waring, Townsend & Lewington, *Field Guide to the Moths of Great Britain and Ireland* (Bloomsbury, 2017), p155
NBDC, *Identification Guide to Ireland's Ladybirds* (2015)
Step, E., *Nature Rambles: Winter to Spring* (1933), p21
Woodward, M., *Gerard's Herbal* (Senate, 1994), p201
Devlin, Z., *Wildflowers of Ireland* (2021), p126, 128
Feehan, J., *Wildflowers of Offaly* (2009), p149, 251

Week 2: Devlin, Z., *Wildflowers of Ireland* (2021), p33
Feehan, J., *Wildflowers of Offaly* (2009), p121
Pratt, A., *Haunts of the Wild Flowers* (1892), p192
Wright, J., *A Natural History of the Hedgerow and ditches, dykes and dry stone walls* (Profile Books, 2017), pp159-163
Feehan, J., Wildflowers of Offaly, YouTube - Blackthorn and Willow
Step, E., *Nature Rambles: Winter to Spring* (1933), pp95-97.
NBDC, *Bumblebee Swatch* (2016)
NBDC, How To Identify and Record Common Irish Bumblebees (biodiversityireland.ie)

Further reading

Goulson, D., *A Sting in the Tale* (Vintage, 2014), pp16-24
Olsen, Sunesen & Pedersen, *Small Woodland Creatures* (Oxford, 2001), pp58-59

Week 3: https://www.mothsireland.com/micro/mages/18003.gif
Sterling, P. & L., *Field Guide to Micro-moths* (2012), p107
Waring, T. & L., *Field Guide to Moths* (2017), p381, 152, 384, 320
Marren, P., *Emperors, Admirals & Chimney Sweepers* (A Little Toller Book, 2020), p192
Lowen, J., *A Gateway Guide British Moths* (Bloomsbury, 2021), p349

Week 4: https://www.naturespot.org.uk/species/water-cricket
Brock, P. D., *Britain's Insects. A field guide to the insects of Great Britain and Ireland* (AES, 2021), p205, 258-9
Olsen, S. & P., *Small Freshwater Creatures* (2001), p96-7, 104, 196
Olsen, S. & P., *Small Woodland Creatures* (2001), p132, 155
Step, E., *Nature Rambles: Winter to Spring* (1933), p39
Step, E., *Nature Rambles: Spring to Summer* (1933), p11-13, 115
Waring, T. & L., *Field Guide to Moths* (2017), p179
Henwood, Sterling & Lewington, *Field Guide to the Caterpillars of Great Britain and Ireland* (Bloomsbury, 2020), p206
South, R., *The Moths of the British Isles* (Warne, 1943) pp62-64
Feehan, J., *Wildflowers of Offaly* (2009), pp386-387, 41
Devlin, Z., *Wildflowers of Ireland* (2021), p216
Step, E., *Nature Rambles: Winter to Spring* (1933), p19, 22
Harding, J. M., *The Irish Butterfly Book* (Butterfly House, 2021), pp70-77, 158-169
NBDC, *Butterfly Swatch* (2017)

April

Week 1: Pratt, A., *Wild Flowers* (1893), vol. i, p27
Pratt, A., *Haunts of the Wild Flowers* (1892), p17

Step, E., *Nature Rambles: Winter to Spring* (1933), pp67-69, 146-147, 81-82

Feehan, J., *Wildflowers of Offaly* (2009), pp4-5, 219-220, 93-97, 450-451, 454-455, 285

NBDC Bank Vole data sheet

Pratt, A., *Wild Flowers* (1893), vol. i, p53

Pratt, A., *Wild Flowers* (1893), vol. ii, p19, 146-147, 27

Devlin, Z., *Wildflowers of Ireland* (2021), pp114-115, 156, 243

Waring, T. & L., *Field Guide to Moths* (2017), p115, 86, 155

Sterling, P. & L., *Field Guide to Micro-moths* (2012), p113

Dempsey, E., *Guide to Ireland's Birds* (2002), p216

Sterry, P., *Collins Guide to British Birds* (Collins, 2004), p256, 252

NBDC, Species Profile, Bank Vole

Week 2: Falk S. & Lewington R., *Field Guide to the Bees of Great Britain and Ireland* (Bloomsbury, 2016), p125, 178, 136, 153, 314, 337

Dunlop, S., (01/06/21), <u>Insects/Invertebrates of Ireland | Facebook</u>

Phillips, R., *Mushrooms* (2006), p367

Sterry, P., *Collins Guide to British Mushrooms* (2009), p304

Olsen, S. & P., *Small Freshwater Creatures* (2001), p121

FSC *A Guide to House and Garden spiders*

Phillips, R., *Grasses. Ferns, Mosses & Lichens* (1980), p26

Step, E., *Nature Rambles: Winter to Spring* (1933), p27

Pratt, A., *Wild Flowers* (1893), vol. i, p23

Feehan, J., *Wildflowers of Offaly* (2009), pp486-488

Week 3: NBDC, *Butterfly Swatch* (2017)

Harding, J. M., *Irish Butterfly Book* (2021), pp94-98, 134-139, 140-146

Marren, P., *Emperors, Admirals & Chimney Sweepers* (2020), p211

Pratt, A., *Wild Flowers* (1893), vol. i, pp155-156

Devlin, Z., *Wildflowers of Ireland* (2021), p233, 39

Feehan, J., *Wildflowers of Offaly* (2009), pp78-79, 240, p244-245

www.ukflymines.co.uk

Further reading

Week 4: Devlin, Z., *Wildflowers of Ireland* (2021), p62
Feehan, J., *Wildflowers of Offaly* (2009), pp468-469, 22-23
Pratt, A., *Haunts of the Wild Flowers* (1892), p28
Ball, S. & Morris R., *Britain's Hoverflies* (Princeton, 2015), p176, 155, 178, 214, 144, 266, 124
Brock, P. D., *Britain's Insects* (2021), pp355-356, 313, 300
NBDC, *Identification Guide to Ireland's Dragonflies and Damselflies* (2013) Swatch
NBDC, *Ladybird Swatch* (2015)
Step, E., *Insect Artizans* (n.d.), p119
Sterling, P. & L., *Field Guide to Micro-moths* (2012), p47

May

Week 1: Chinery, M., *Garden Wildlife* (1997), p14
www.irishhedgehogsurvey.com
Step, E., *Animal Life of the British Isles* (1942), pp9-14, 78-81, 113-117
Dempsey, E., *Guide to Ireland's Birds* (2002), p146
Pratt, A., *Haunts of the Wild Flowers* (1892), p236
Harrap, S., *Harrap's Wild Flowers* (Bloomsbury Wildlife, 2013), p162, 247, 48-49
Pratt, A., *Wild Flowers* (1893), vol. ii, p10, 71-72
Feehan, J., *Wildflowers of Offaly* (2009), p363, 340
Pratt, A., *Wild Flowers* (1893), vol. i, p87
NBDC (2023) Spiders of Ireland (biodiversityireland.ie)
Olsen, S. & P., *Small Woodland Creatures* (2001), p89, 64
Falk & L., *Field Guide to Bees* (2016), p150, 162, 217, 413
Beckett, O., (2023) The Irish Naturalist, Forest Cuckoo Bumblebee (Bombus sylvestris) - The Irish Naturalist

Week 2: NBDC, *Butterfly Swatch* (2017)
Harding, J. M., *Irish Butterfly Book* (2021), pp85-88, 52-58, 59-64, 46-51, 115-120, 99-102, 127-133

Devlin, Z., *Wildflowers of Ireland* (2021), p132, 184, 205, 111
Pratt, A., *Wild Flowers* (1893), vol. ii, p50
Feehan, J., *Wildflowers of Offaly* (2009), p379, 99
Brock, P. D., *Britain's Insects* (2021), p344
Olsen, S. & P., *Small Woodland Creatures* (2001), p112
Dunlop, S., (21/05/20), Insects/Invertebrates of Ireland | Facebook
www.sawflies.org.uk/euura-pavida
Olsen, S. & P., *Small Freshwater Creatures* (2001), p36
Nelson, B. & Thompson, R., *The Natural History of Ireland's Dragonflies* (Ulster Museum, 2004), pp4-14
Brooks, S., & Lewington, R., *Field Guide to the Dragonflies and Damselflies of Great Britain and Ireland* (British Wildlife, 1997), p15
NBDC, *Dragonflies and Damselflies Swatch* (2013)
Waring, T. & L., *Field Guide to Moths* (2017), p57, 212, 218
Sterling, P. & L., *Field Guide to Micro-moths* (2012), p379
Dempsey, E., *Guide to Ireland's Birds* (2002), p154, 152
www.birdwatchireland.ie, *Saving Swifts Guidelines*
Sterry, P., *Collins Guide to British Birds* (2004), p174
Johnson, W., *Gilbert White's Journals* (Routledge, 1970), p410

Week 3: Feehan, J., *Wildflowers of Offaly* (2009), p242, 291, 377-378
Devlin, Z., *Wildflowers of Ireland* (2021), p201, 104, 116, 161, 135
Chinery, M., *Britain's Plant Galls* (Wild Guides, 2016), p81, 85
Brock, P. D., *Britain's Insects* (2021), p284, 245, 195, 180, 219
Olsen, S. & P., *Small Woodland Creatures* (2001), p101, 128
Southwood & Leston, *Land & Water Bugs of the British Isles* (Warne, 1959), p277, 305, 215, 86
Ball & M., *Britain's Hoverflies* (2015), p112
www.sawflies.org.uk/
Olsen, S. & P., *Small Freshwater Creatures* (2001), p52, 55

Week 4: Sterry, P., *Collins Guide to British Mushrooms* (2009), p316
Devlin, Z., *Wildflowers of Ireland* (2021), p37

Further reading

Feehan, J., *Wildflowers of Offaly* (2009), p160
Olsen, S. & P., *Small Woodland Creatures* (2001), p53, 38, 93
Brock, P. D., *Britain's Insects* (2021), p481, 128
NBDC, *Shieldbugs Swatch* (2012)
Southwood & L., *Land & Water Bugs* (1959), p48
Waring, T. & L., *Field Guide to Moths* (2017), p50
South, R., *Moths of the British Isles* (1943) pp121-123
Sterling, P. & L., *Field Guide to Micro-moths* (2012), p70
Marren, P., *Emperors, Admirals & Chimney Sweepers* (2020), p117
NBDC, *Butterfly Swatch* (2017)
Harding, J. M., *Irish Butterfly Book* (2021), p191-198

June

Week 1: NBDC, *Identification Guide to Ireland's Shieldbugs* (2012) Swatch
www.ukbeetles.co.uk
Brock, P. D., *Britain's Insects* (2021), p316, 196, 517
Olsen, S. & P., *Small Woodland Creatures* (2001), p107, 110
www.britishbugs.org.uk
Ball & M., *Britain's Hoverflies* (2015), p138
NBDC Species Profile, Wasps
Feehan, J., *Wildflowers of Offaly* (2009), p390-391, 332, 458-459
Waring, T. & L., *Field Guide to Moths* (2017), p55, 26, 43, 335, 303, 39
Henwood, S. & L., *Field Guide to Caterpillars* (2020), p128
Sterling, P. & L., *Field Guide to Micro-moths* (2012), p106, 202

Week 2: Harding, J. M., *Irish Butterfly Book* (2021), pp234-239
Chinery, M., *Britain's Plant Galls* (2016), p56
Ball & M., *Britain's Hoverflies* (2015), p98, 122
Devlin, Z., *Wildflowers of Ireland* (2021), p291, 293, 226, 318
Feehan, J., *Wildflowers of Offaly* (2009), p452, 449, 457
www.sawflies.org.uk

Bernard, P., & Ross, E., *The Adult Trichoptera (Caddisflies) of Britain and Ireland* (British Entomological Society, 2012), p128, 138, 121, 174, 178, 102, 143, 98

O'Connor, J. P., *Atlas of the Irish Trichoptera (Caddisflies)* (Irish Biogeographical Society, 2021), with maps for each species p112, 99, 96, 151, 153-157, 79, 130

Olsen, S. & P., *Small Freshwater Creatures* (2001), p32, 19

Brock, P. D., *Britain's Insects* (2021), p371, 145, 237

Week 3: Feehan, J., *Wildflowers of Offaly* (2009), p169, 298, 54

Step, E., *Wayside and Woodland Ferns* (Warne & Co., 1908), pp91-94

Devlin, Z., *Wildflowers of Ireland* (2021), p163, 168, 199, 181

Harrap, S., *Harrap's Wild Flowers* (2013), p352, 292, 233, 185

Waring, T. & L., *Field Guide to Moths* (2017), p380, 138

Henwood, S. & L., *Field Guide to Caterpillars* (2020), p337, 180

Harding, J. M., *Irish Butterfly Book* (2021), pp230-233, 78-83, 225-229, 147-151

Dunlop, S., (2023), Featured: White Butterflies, Insects/Invertebrates of Ireland | Facebook

Pedlow, A., *Did you Know? 100 quirky facts about County Offaly* (OCC, 2013), pp46-47

Brock, P. D., *Britain's Insects* (2021), p108

Olsen, S. & P., *Small Woodland Creatures* (2001), pp98-99

Southwood & L., *Land & Water Bugs* (1959), p241, 239, 311, 299

Britishbugs.org.uk (Miridae)

Britishbugs.org.uk (Delphacidae)

Step, E., *Animal Life of the British Isles* (1942), pp59-69

National Parks and Wildlife Service, (2014) pine-marten-in-houses.pdf (npws.ie) and (2008) Otter Leaflet 4 (npws.ie)

Week 4: Waring, T. & L., *Field Guide to Moths* (2017), p151, 167, 413, 298, 147, 173, 294, 388, 391, 202, 62, 55, 198

Henwood, S. & L., *Field Guide to Caterpillars* (2020), p344, 284, 131, 128, 282

Further reading

Ball & M., *Britain's Hoverflies* (2015), p270, 244, 204
NBDC, *Dragonflies and Damselflies Swatch* (2013)
Brooks & L., *Dragonflies and Damselflies* (1997), pp62-63
Thompson, R., & Nelson, B., *Guide to the Dragonflies and Damselflies of Ireland* (National Museums Northern Ireland, 2014), pp16-17
Olsen, S. & P., *Small Woodland Creatures* (2001), p124, 134
Brock, P. D., *Britain's Insects* (2021), p301, 285

July

Week 1: NBDC, *Dragonflies and Damselflies Swatch* (2013)
Olsen, S. & P., *Small Woodland Creatures* (2001), p96
NBDC, Species Profile, *Ectemnius lapidarius*
Falk & L., *Field Guide to Bees* (2016), p384
Ball & M., *Britain's Hoverflies* (2015), p114, 246
Feehan, J., *Wildflowers of Offaly* (2009), pp266-267, 381, 335, 399-400
Pratt, A., *Haunts of the Wild Flowers* (1892), p64

Week 2: Feehan, J., *Wildflowers of Offaly* (2009), p185
Pratt, A., *Wild Flowers* (1893), vol. ii, p257
Southwood & L., *Land & Water Bugs* (1959), p239, 45
Devlin, Z., *Wildflowers of Ireland* (2021), p24, 236
Olsen, S. & P., *Small Woodland Creatures* (2001), pp8-9, 68
Sterling, P. & L., *Field Guide to Micro-moths* (2012), p376-377
www.leafmines.co.uk
NBDC, *Shieldbugs Swatch* (2012)

Week 3: Waring, T. & L., *Field Guide to Moths* (2017), p146, 413, 187
South, R., *Moths of the British Isles* (1943) vol. ii, p284, vol. i, p224, 84
Sterling, P. & L., *Field Guide to Micro-moths* (2012), p57
Langmaid, P. & Y., Field *Guide to Smaller Moths* (2018), p42
Feehan, J., *Wildflowers of Offaly* (2009), p306, 336, 433, 430

Dempsey, E., *Guide to Ireland's Birds* (2002), p202
www.birdwatchireland.ie
Brock, P. D., *Britain's Insects* (2021), p335, 290-295
Olsen, S. & P., *Small Woodland Creatures* (2001), p118

Week 4: Sterry, P., *Collins Guide to British Mushrooms* (2009), p302
Phillips, R., *Mushrooms* (2006), p374
Waring, T. & L., *Field Guide to Moths* (2017), p313, 148
Sterling, P. & L., *Field Guide to Micro-moths* (2012), p385, 78, 102, 262
Step, E., *Wayside and Woodland Ferns* (1908), pp51-52
Langmaid, JR., Palmer, SM., Young, MR., *A FieldGuide to the Smaller Moths of Great Britain and Ireland* (British Entomological and Natural History Society, 2018), p61
Devlin, Z., *Wildflowers of Ireland* (2021), p273
Feehan, J., *Farming in Ireland, History, Heritage and Environment* (UCD, 2003), pp435-437
FSC *A Guide to House and Garden spiders*

August

Week 1: Ball & M., *Britain's Hoverflies* (2015), p242, 110
NBDC, Species Profile, *Sphaerophoria scripta*
Mirids—www.britishbugs.org.uk
Southwood & L., *Land & Water Bugs* (1959), p283
Chinery, M., *Britain's Plant Galls* (2016), p90
www.sawflies.org.uk
Feehan, J., *Wildflowers of Offaly* (2009), p247, 432, 49
Waring, T. & L., *Field Guide to Moths* (2017), p340, 336, 170, 302

Week2: Harding, J. M., *Irish Butterfly Book* (2021), pp185-190
Nash, D., Boyd, T., Hardiman, D., *Ireland's Butterflies A Review*, (The Dublin Naturalist's Field Club, 2012), pp183-185
Falk & L., *Field Guide to Bees* (2016), p306, 311-312, 293, 304

Further reading

NBDC, Species Profile, *Coelioxys elongata* and *Coelioxys inermis*
www.britishbugs.org.uk
Brock, P. D., *Britain's Insects* (2021), p215
Chinery, M., *Britain's Plant Galls* (2016), p59, 32, 62, 55
Nelson, & Thompson, *Ireland's Dragonflies* (2004), pp144-151
Feehan, J., *Wildflowers of Offaly* (2009), pp292-293, 334, 50
NBDC *The secret life of solitary bees*
Olsen, S. & P., *Small Woodland Creatures* (2001), p56

Week 3: NBDC, Species Profile, Common Green Grasshopper
Olsen, S. & P., *Small Freshwater Creatures* (2001), p112
Brock, P. D., *Britain's Insects* (2021), p471
Olsen, S. & P., *Small Woodland Creatures* (2001), pp49-50
Belfast Hills Partnership
NBDC, *Butterfly Swatch* (2017)
McCormack S. & Regan E., *Insects of Ireland, A Field Guide* (Collins Press, 2016), p75, 69
Southwood & L., *Land & Water Bugs* (1959), p107, 269
www.britishbugs.org.uk
Sterling, P. & L., *Field Guide to Micro-moths* (2012), p194

Week 4: Devlin, Z., *Wildflowers of Ireland* (2021), p14
Southwood & L., *Land & Water Bugs* (1959), p128, 240, 278, 205
Brock, P. D., *Britain's Insects* (2021), p184
Ball & M., *Britain's Hoverflies* (2015), p126
www.britishbugs.org.uk
Nelson, & Thompson, *Ireland's Dragonflies* (2004), pp183-190
Thompson, & N., *Dragonflies and Damselflies of Ireland* (2014), pp66-73
Feehan, J., *Wildflowers of Offaly* (2009), pp235-236, 464
www.sawflies.org.uk
Falk & L., *Field Guide to Bees* (2016), p79

Feehan, J., *Wildflowers of Offaly* (2009), pp198-199
Smart, B., *Micro-moth FieldTips* (Lancashire & Cheshire Fauna Society, 2018), vol. ii, p138
Sterling, P. & L., *Field Guide to Micro-moths* (2012), p152
www.ukmoths.org.uk

September

Week 1: South, R., *Moths of the British Isles* (1943) vol. i, p20
Waring, T. & L., *Field Guide to Moths* (2017), p359
Devlin, Z., *Wildflowers of Ireland* (2021), p96, 27, 164, 151
Feehan, J., *Wildflowers of Offaly* (2009), p189, 20, 411-412, 248
Pratt, A., *Wild Flowers* (1893), vol. i, p55
Southwood & L., *Land & Water Bugs* (1959), p300
Brock, P. D., *Britain's Insects* (2021), p199, 105, 114
NBDC Species Profile, *Platycheirus granditarsus*
Ball & M., *Britain's Hoverflies* (2015), p88
NBDC Species Profile, *Mellinus arvensis*
Falk & L., *Field Guide to Bees* (2016), p92, 142
FSC *A Guide to House and Garden spiders*
Olsen, S. & P., *Small Freshwater Creatures* (2001), p91
NBDC, *Dragonflies and Damselflies Swatch* (2013)
Nelson, & Thompson, *Ireland's Dragonflies* (2004), pp286-293

Week 2: Southwood & L., *Land & Water Bugs* (1959), p273, 292. 290, 230, 229
www.britishbugs.org.uk
Pratt, A., *Wild Flowers* (1893), vol. i, p109
Devlin, Z., *Wildflowers of Ireland* (2021), p48
Feehan, J., *Wildflowers of Offaly* (2009), p218
Chinery, M., *Britain's Plant Galls* (2016), p162, 35, p42, p75
www.sawflies.org.uk
www.leafmines.co.uk

Further reading

Week 3: Harding, J. M., *Irish Butterfly Book* (2021), pp65-68
Nash, B. & H., *Ireland's Butterflies A Review*, (2012), pp123-125
Waring, T. & L., *Field Guide to Moths* (2017), p77, 195, 377
South, R., *Moths of the British Isles* (1943) vol. ii, pp141-142
Southwood & L., *Land & Water Bugs* (1959), pp291-292
Ball & M., *Britain's Hoverflies* (2015), p206
Olsen, S. & P., *Small Woodland Creatures* (2001), p99, 163
Brown, V. K., *Grasshoppers* (Naturalists' Handbook 2, 1990), p30-31
Chinery, M., *Britain's Plant Galls* (2016), p25, 28, 27
Redfern, M., & Askew, R. R., *Plant Galls* (Naturalists' Handbook 17, 1992), p60, 59
Sterry, P., *Collins Guide to British Mushrooms* (2009), p274, 244, 220, 200, 308, 80
Phillips, R., *Mushrooms* (2006), p338, 260, 231, 377

Week 4: www.birdwatchireland.ie
Sterry, P., *Collins Guide to British Birds* (2004), p184, 188
Dempsey, E., *Guide to Ireland's Birds* (2002), p166, 178
www.leafmines.co.uk
Sterry, P., *Collins Guide to British Mushrooms* (2009), p90, 218, 124
Phillips, R., *Mushrooms* (2006), p140, 350, 257, 102
Dunlop, S., (20/09/21), Insects/Invertebrates of Ireland | Facebook
Bernard & Ross, *Adult Trichoptera (Caddisflies)* (2012), p147
Brock, P. D., *Britain's Insects* (2021), p243
www.ukbeetles.co.uk

October

Week 1: Waring, T. & L., *Field Guide to Moths* (2017), p58, 319
Ward, M., *Sketches with the Microscope in a Letter to a Friend, A reproduction with essays by Michael Byrne and John Feehan* (Offaly Historical and Archaeological Society, 2019), p14
Sterry, P., *Collins Guide to British Mushrooms* (2009), p270, 310

Feehan, J., *Wildflowers of Offaly* (2009), p286
Johns, Rev. C. A., *Botanical Rambles* (London, c.1850), pp9-12

Week 2: Waring, T. & L., *Field Guide to Moths* (2017), p306, 363, 366, 361, 368
Phillips, R., *Grasses. Ferns, Mosses & Lichens* (1980), p154, 153, 155, 122, 142

Week 3: Sterry, P., *Collins Guide to British Mushrooms* (2009), p30, 152
www.leafmines.co.uk
www.britishbugs.org.uk
Southwood & L., *Land & Water Bugs* (1959), p161, 153
Olsen, S. & P., *Small Woodland Creatures* (2001), p100
Harding, J. M., *Irish Butterfly Book* (2021), pp199-204

Week 4: Sterry, P., *Collins Guide to British Mushrooms* (2009), p106, 272, 274, 335
Phillips, R., *Mushrooms* (2006), p248

November

Week 1: Step, E., *Animal Life of the British Isles* (1942), p130
NBDC, Species Profile, Fallow Deer
Chinery, M., *Britain's Plant Galls* (2016), p17, 16, 21
Redfern, & A., *Plant Galls* (1992), p62, 64, 67
Dempsey, E., *Guide to Ireland's Birds* (2002), p114
www.birdwatchireland.ie
Sterry, P., *Collins Guide to British Birds* (2004), p130
Sterry, P., *Collins Guide to British Mushrooms* (2009), p306, 308, 316, 240
Phillips, R., *Mushrooms* (2006), p376, 346
Feehan, J., *Wildflowers of Offaly* (2009), p397

Further reading

Week 2: Brock, P. D., *Britain's Insects* (2021), p473, 309
South, R., *Moths of the British Isles* (1943) vol. i, p288, 279
Waring, T. & L., *Field Guide to Moths* (2017), pp318-319, 151-2
Henwood, S. & L., *Field Guide to Caterpillars* (2020), p189
Dempsey, E., *Guide to Ireland's Birds* (2002), p40
Sterry, P., *Collins Guide to British Mushrooms* (2009), p102, 194, 136, 124

Week 3: Langmaid, P. & Y., *Field Guide to Smaller Moths* (2018), p36
Sterling, P. & L., *Field Guide to Micro-moths* (2012), p53
Sterry, P., *Collins Guide to British Birds* (2004), p82
Dempsey, E., *Guide to Ireland's Birds* (2002), pp72-73

Week 4: Feeley, H., (16/11/23) *Secrets of the Stream, Ireland's Stoneflies*, Webinar, Cork Nature Network
Feeley, H., Baars, J-R., Kelly-Quinn, M., *The Stonefly (Plecoptera) of Ireland* (NBDC, 2016), pp40-41, 66-67, 56-57
Waring, T. & L., *Field Guide to Moths* (2017), p363
South, R., *Moths of the British Isles* (1943) vol. i, p14
Sterling, P. & L., *Field Guide to Micro-moths* (2012), p164
Sterry, P., *Collins Guide to British Birds* (2004), p110, 112, 54, 102, 196
www.birdwatchireland.ie

December

Week 1: Sterry, P., *Collins Guide to British Mushrooms* (2009), p304, 316
Feehan, John, *Summer School notes*, (Offaly County Council) unpublished
Sterry, P., *Collins Guide to British Birds* (2004), p220, 224, 252, 190, 246

Week 2: Olsen, S. & P., *Small Woodland Creatures* (2001), pp54-56, 182, 140, 142

NBDC, *Ladybird Swatch* (2015)
McCormack S. & R., *Insects of Ireland, A Field Guide* (2016), p71, 134-136
Chinery, M., *Garden Wildlife* (1997), p56, 182
Ward, M., *Sketches with the Microscope* (2019), p2

Week 3: Southwood & L., *Land & Water Bugs* (1959), p80
www.britishspiders.org.uk
Chinery, M., *Garden Wildlife* (1997), p164
Step, E., *Nature Rambles: Winter to Spring* (1933), pp56-58
Brock, P. D., *Britain's Insects* (2021), p340

Week 4: Sterry, P., *Collins Guide to British Birds* (2004), p236, 240, 238
Hatch, N., *Ireland's Crows, A beginner's Guide to Ireland's Birds* (Birdwatch Ireland, 2020), Wings Autumn
Dempsey, E., *Guide to Ireland's Birds* (2002), p206, 208
Waring, T. & L., *Field Guide to Moths* (2017), p100, 47
Henwood, S. & L., *Field Guide to Caterpillars* (2020), p156

Grass-of-Parnassus,
Parnassia palustris

Index

10-spot Ladybird	381	*Apolygus spinolae*	256
14-spot Ladybird	**41, 211**, 278, 381	Apple	177, 331, **332**, 336
22-spot Ladybird	381	*Arctosa perita*	**298-299**
2-spot Ladybird	278	*Arge ustulata*	281
7-spot Ladybird	**24, 41**	*Argyresthia brockeela*	230
Abia nitens	173	*Argyresthia goedartella*	230
Acalitus longisetosus	149	Ashy Mining Bee	**83-84**, 86
Acrolepia autumnitella	344	Aspen	66, 172, 214-215, 288, 366-367
Adelphocoris lineolatus (Lucerne Bug)	302	Aubretia	49, 330
Adelphocoris seticornis	**314-315**	Autumn Gentian	**270-271**
Agaricus campestris	326	Autumn Lady's-tresses	287
Aglaostigma aucupariae	155	Azure Damselfly	142
Alder	82, 138, 154, 173, 186-187, 189, 208, 30, 234, 248, 266-267, 296, 308-309	**Bagworm**	**166-167**
		Ballinagar	117
Alder Tongue	**266-267**	Ballykealy	124
Alderfly	193	Ballylin Bog	303, 331, 344
Alderfly, *Sialis* sp.	193	Banagher	70, 374
Alexanders	43	Banded Demoiselle	**212-213**
Allantus sp.	**188-189**	Bank Vole	76
Amber Snail	386	Barn Owl	322
Amethyst Deceiver	**364-365**	Barren Strawberry	38
Ampedus sp.	243	Beautiful China-mark	**232-233**
Ancistrocerus gazella	**162-163**	Beautiful Demoiselle	213
Ancistrocerus trifasciatus	171	Bee Orchid	175
Andrena praecox	128	Bee	9, 40, 46, **49-53**, 54, 74, 78-79, **83-87**, 95, 103, 107, 108, 125, 127, **128-129**, 148, 163, 184, 186, 210, **222**, **224-225**, 252-253, 258, **264**, **272-273**, 287, **289**, 290, 295, **297**, 314-315, 334
Andrena sp.	**83-86, 128**, 297		
Andrena wilkella	128		
Andricus curvator	355		
Andricus kollari	355		
Andricus quercuscalicis	355		
Angle Shades	178	**Bee,** *Andrena praecox*	128
Ant	64-65, 98, 222, 315, 380	*Andrena* sp.	**83-86, 128**, 297
Formica sp.	380	*Andrena wilkella*	128
Lasius sp.	380	Blood Bee	128
Myrmica sp.	380	*Bombus* sp.	6-7, **49-53**, 129, 210, **224**, 231
Aphid	30, 41, 108, 110, 113, 163, 170-171, 183, 254, 276, 278, **328**, 345, 380-381	Bumblebee, Buff-tailed	**49-50**
		Bumblebee, Early	**52-53**, 129
		Bumblebee, Garden	**52-53**, **224**
Aphid gall	**266-267**, **306-307**	Bumblebee, Red-tailed	9, **52-53**, 210, 222

410

Step into Nature

Bee, Bumblebee, White-tailed **49-50**, 210
    Coelioxys sp.   **264**, 273
    Colletes succinctus   **289**
    Common Carder Bee   **51-53**, 210
    Four Coloured (Forest) Cuckoo Bee **129**
    Megachile (leafcutter)   84, 264, **272-273**
    Mining, Ashy   **83-84**, 86
    Mining, Chocolate   **85**
    Mining, Early   **84-85**
    Nomada goodeniana   **85-86**
    Nomada marshamella   **85-86**
    Orange-legged Furrow Bee **84**, 128
    Sharp-tail Bee   **264**, 273
    Small Scabious Mining Bee   **297**
Beech Jellydisc   **369**
Beetle   **32**, **58-59**, **64-65**, 69, 74, 103, **114-115**, 116, **137**, 146, **150-151**, 161, **163**, **169-170**, **174**, **216-217**, **243**, 258, **275**, 278, 328, 334

Beetle, *Ampedus* sp.   **243**
    Cardinal, Black-headed   151
    Cardinal, Red-headed   **150-151**
    Cassida rubiginosa   **169**
    Cassida vibex   **169**
    Click Beetle   **243**
    Common Cockchafer (May Bug)   **150-151**
    Ctenicera cuprea   **243**
    Diving, Large-grooved   59
    Diving, Lesser   **58-59**
    Donacia sp.   **275**
    Garden Chafer   **217**
    Green Tiger Beetle   **64-65**
    Longhorn, Four-banded   **216-217**
    Longhorn, Lesser Thorn-tipped **170**
    Longhorn, Two-banded   **114-115**
    Malachite   **150-151**
    Malthinus flaveolus   **163**
    Malthodes marginatus   **163**
    Pill, Common   **150-151**
    Sphaeridium sp.   **328-329**
    Weevil, Birch Leaf Roller   **137**
    Weevil, Figwort   **174**
    Weevil, Gorse   **32**
Berries   40, 82, 92, 100, 204, 292, 313, 337, 373-374, 388
Bilberry   **100**
Birch   41, 82, 87, 112, 115-116, 118, 120, 136-137, 139, 149, 154, 161, 163, 173, 182, 189, 206, 217, 230, 237, 242, 246-248, 278, 280-281, 283, 292, 296, 308-309, 313, 326, 341, 366
Birch Bolete   **342-343**
Birch Catkin Bug   **154**
Birch Leaf Roller   **137**
Birch Shieldbug   257, **341**
Bird, Barn Owl   322
    Blue Tit   242, **378**
    Buzzard   **368**, 388
    Coal Tit   234, **378**, **393**
    Cough   389
    Crow   118, 204, **388-389**
    Cuckoo   96, 118, 221
    Dunnock   **379**
    Eagle   368
    Bird, general   24, 28-29, 40, 50, 82, 144-145, 173, 177, 242, 322-323, 333, 374-375, 378-379, 388-389
    Golden Plover   **374**
    Goldfinch   **379**
    Great Tit   **378**
    Greenfinch   9, **378-379**
    Heron, Grey   **363**
    Hooded Crow   **389**
    House Martin   **145**

Index

Bird, Jackdaw — **389**
Jay — 389
Kestrel — 368
Lapwing — **374**
Magpie — 389
Long-tailed Tit — **28-29**
Mallard — **374-375**
Moorhen — 375
Raven — **388**
Rook — **388-389**
Siskin — **82**
Snipe — **356**
Sparrow, House — **18**, 379
Starling — **40**, 292
Stonechat — 375
Swallow — **144-145**
Swift — **144-145**
Treecreeper — 242
Wagtail, Grey — 323
Wagtail, Pied — **377**, 378
Whooper Swan — **30**
Wren — **379**
Yellowhammer — **82**
Bird's-foot-trefoil — 128, 132, **134**, 143, 179, 244, 275, 310, 344
BirdWatch Ireland — 322, 356
Birr Demesne — 26, 43, 75, 117, 154, 172, 179, 196, 205, 210, 212, 254, 333, 344-346
Black Darter — **300**
Black Rustic — **313**
Blackberry — 332
Black-tailed Skimmer — **220**
Blackthorn — **46-48**, 83, 85, 177, 181, 257, 313, 325, 331, **332**, 336
Bladderwort — **294-295**
Blair's Shoulder-knot — **336-337**
Blood Bee — **128**
Bloody Crane's-bill — **159**

Blue Fleabane — **251**, 310
Blue Roundhead — **350-351**
Blue Shieldbug — **168**
Blue Tit — 242, **378**
Bluebell — 70, 74, **104-106**, 108
Blue-tailed Damselfly — **142**, 269
Bog Asphodel — **194-195**
Bog Beacon — **88**
Bog Pimpernel — **196-197**
Bog Rosemary — **100**
Bog Stitchwort — **124**
Bogbean — **124**
Bombus sp. — 6-7, **49-53**, **129**, 210, **224**, 231
Bordered Beauty — **236**
Bracken — 188-189, 203, 307
Brambles — 188, 199, 256, 263
Brick — **373**
Brimstone — **70-71**
Broad Damselbug — **345**
Bronze Shieldbug — **161**, **341**
Broom — 80, 118, 121, 203
Broom Fork Moss — **338-339**
Brown China-mark — **233**
Brown Hawker — **218-219**
Brown Lipped Snail — **386**
Buckthorn — 70, 98, 163, **190-191**
Buff Ermine — **276**
Buff-tailed Bumblebee — **49-50**
Bug — 24, 30-32, 44, 58, 74, 103, **151-154**, **160-161**, 168, 170, **202-203**, 221, **226-227**, **234**, **244-245**, **256-257**, 265, **279**, **282**, **284**, **296**, **302-303**, **304**, **341**, **384-385**
Capsus ater — **202-203**
Chilacis typhae — **384-385**
Damselbug, Broad — 345
Damselbug, Marsh — 345
Ditropis pteridis — 203
Drymus sylvaticus — **278-279**

Bug, leafhopper	265, 279	Burnet Companion	**143**
Cicadella viridis	**265**	Burnet Rose	**159**
Evacanthus interruptus	**265**	Burnished Brass	**206**
Zonocyba bifasciata	**265**	Bur-reed	232-233
Bug, *Metatropis rufescens*	**282**	Burren 159, 179, 190, 204, 208, 243, 272, 287, 298	
Bug, mirid, *Adelphocoris lineolatus* (Lucerne Bug)	**302**	Bush Vetch	125, **332**, 344
		Buttercup	27, 115
Adelphocoris seticornis	**314-315**	**Butterfly** 63, 46-47, 54, 69, **70-73**, **94-99**, 101,	
Apolygus spinolae	**256**	103, 125, **132-133**, 134, 167, **180-181**,	
Birch Catkin Bug	**154**	**199-200**, 241, 258, **262-263**, 270, 277,	
Calocoris roseomaculatus	**244-245**	**310-311**, 331, 334, **346-347**	
Campyloneura virgula	**202**	Blue, Common	98, **133**
Deraeocoris lutescens	**284**	Blue, Holly	**98-99**
Dichypus epilobii	**227**	Blue, Small	**98**
Dicyphus errans	**202**	Brimstone	**70-71**
Dicyphus pallicornis	**284**	Clouded Yellow	**310-311**, 312
Dicyphus stachydis	**284**	Comma	**72-73**, 95
Grypocoris stysi	**170**	Dingy Skipper	**132**
Harpocera thoracica	**154**	Fritillary, Silver-washed	**262-263**
Heterocordylus tibialis	**203**	Green Hairstreak	**133**
Heterotoma planicornis	**152-153**	Meadow Brown	**199**
Leptopterna dolabrata	**202**	Orange-tip	**94-97**
Liocoris tripustulatus	**152-153**	Painted Lady	**200**
Lucerne Plant Bug	**302**	Peacock	47, **71**, 72, 241
Lygus wagneri	**302**	Red Admiral	72, **99**
Malacocoris chlorizans	**279**	Ringlet	**199**
Pantilius tunicatus	**296**	Small Copper	**133**
Pinalitus cervinus	**284**	Small Heath	**180-181**
Pithanus maerkeli	**279**	Small Tortoiseshell	**71-72**
Plagiognathus arbustorum	**303**	Speckled Wood	**346-347**
Plagiognathus chrysanthemi	**303**	White, Green-veined	43, **96**, 241
Potato Capsid Bug	**303**	White, Large	**199**, **277**
Stenodema laevigata	**152-153**	White, Small	101, **132**, 199
Bug, Scale Insect	**163**	White, Wood	**132**
Thread-legged Bug	**345**	Buzzard	**368**, **388**
Water Cricket	**58**	**Caddisfly**	37 (case), **190-192**, 323, 328
Water Measurer	**62**	*Glyphotaelius pellucidus*	**190-191**
Bulrush (Reedmace)	384-385	*Hagenella clathrata*	**192**

Index

Caddisfly, *Halesus radiatus* **328**
   Limnephilus auricula **190**
   Limnephilus lunatus **190**
   Limnephilus sp. **37**
   Mystacides azurea **190-191**
   Oecetis ochracea **191**
   Phryganea sp. **192**
   Stenophylax permistus **192**
Caliroa cerasi **257**
Callows 184
Calocoris roseomaculatus **244-245**
Camcor River, Birr 323, 370
Campodorus sp. **340**
Campyloneura virgula **202**
Canary-shouldered Thorn **248-249**
Candlesnuff Fungus **18-19**
Candy-striped Spider **283**
Capsus ater **202-203**
Cassida rubiginosa **169**
Cassida vibex **169**
Cat's Paw (Mountain Everlasting) **146**
Celandine, Lesser **26-27**, 74, 106
Ceratiomyxa fruticulosa **352-353**
Chalcid wasp 253
Charleville Demesne 104, 106, 315, 369, 383
Chickweed, Common **22-23**, 45, 124
Chilacis typhae **384-385**
Chinese Character **177**
Chirosia grossicauda **307**
Chocolate Mining Bee **85**
Chough 389
Chrysotoxum bicinctum **182**
Cicadella viridis **265**
Cimbex connatus **186-187**
Cimbex femoratus **139**, 186
Cimbex luteus 186
Cladius brullei **309**
Cladius grandis **288**

Clara Bog, Esker and boardwalk 69-70, 72, 100-101, 165, 168, 185, 265, 280, 301
Clematis 53, 294-295
Click Beetle 243
Cloeon dipterum **156**
Cloghan Lakes 112, 196, 231, 251
Clonfinlough 146, 286
Clonmacnoise 94, 143, 148, 196, 339, 386
Clooneen Bog 101, 113, 139, 143, 154, 160, 169, 189, 201, 220, 252, 260, 285, 298, 342
Clouded Drab **198**
Clouded Funnel **350-351**
Clouded Yellow **310-311**, 312
Clover 128, 244, **290-291**, 302, 310-311, 313
Coal Tit 234, **378**
Cochylimorpha straminea **252-253**
Coelioxys sp. **264**, 273
Coleophora deauratella **290-291**
Collared Earthstar **333**
Colletes succinctus **289**
Coltsfoot **26-27**, 42
Comma **72-73**, 95
Common Blue 98, **133**
Common Blue Damselfly **140**, 142
Common Butterwort **124-125**, 295
Common Carder Bee **51-53**, 210
Common Centaury **222**, 310
Common Chickweed **22-23**, 45, 124
Common Cockchafer (May Bug) **150-151**
Common Cottongrass **260**
Common Darter **220**
Common Earwig **382-383**
Common Field-speedwell **42**
Common Figwort **174**
Common Fleabane **294-295**
Common Frog 24, **36**, 40, 89, 205, 322, 363, 368, 386

Common Froghopper	221	Crystal Brain	**369**
Common Green Grasshopper	**274-275**, 201	*Ctenicera cuprea*	**243**
Common Groundhopper	301	Cuckoo	96, 118, 221
Common Hawker	285	Cuckooflower (Lady's Smock)	**94-97**, 150, 155
Common Knapweed	6-7, 9, 199, **252-253**, 254, 310	Cuckoo-spit	221
Common Mouse-ear	45	Cup Lichen	**33**
Common Pill Beetle	**150-151**	**Daingean**	115
Common Puffball	**318**, 333, 348	Daisy	27, 79, **334-335**
Common Purple and Gold	143	Dame's Violet	**54-55**, 178
Common Quaker	**56-57**	Damselfly, Blue-tailed	**142**, 269
Common Spangle Gall	317	**Damselfly**	111, 140, **142**, 156, **212-213**, 218, **268-269**
Common Twayblade	79	Azure	**142**
Common Valerian	**196-197**	Common Blue	**140, 142**
Common Vetch	125	Demoiselle, Banded	**212-213**
Common Whitlowgrass	45	Demoiselle, Beautiful	**213**
Conopid, *Myopa* sp.	**127**	Emerald	**218**
Sicus ferrugineus	**127**, 252-253	Large Red	**111**
Cooleeny	76, **78-79**	Scarce Blue-tailed	**268-269**
Coral Spot	**358**	Variable	**142**
Cotesia glomeratus	**277**	Dandelion	27, 50, 71, 85-86, 95, 129, 175
Cow Parsley	94, **102-103**	Dark Honey Fungus	**348**
Cowslip	**77-78**	Dark Tussock	119
Cow-wheat	**222-223**	Darwin, Charles	50, 185
Crab Apple	331, **332**, 336	*Dasineura ulmaria*	149
Cranberry Bog, Gallen, (part of Noggusboy Bog) 31-32, **62**, 65, 110-115, 120-121, 124, 132, 139, 141, 143, 150, 172, 190, 193, 203, 205, 209, 214-215, 218-219, 241, 243, 245, 251, 256, 264, 266, 271, 278, 280-281, 289, 292, 301, 304, 333, 341, 344, 384		Deceiver	**327, 350-351**
		December Moth	**390-391**
		Deraeocoris lutescens	**284**
		Derryounce Lake	310-311, 313
		Devil's Matchstick	**33**
Cranefly, *Tanyptera atrata*	**136**	Devil's-bit Scabious	167, 173, 237, **254-255**, 297
Cranesbill	39, 51, 89, **159**	Diamond-back	55, **178**
Cream-spot Ladybird	**113**	*Dichypus epilobii*	**227**
Creeping Cinquefoil	**147**	*Dicyphus errans*	**202**
Crinkill	44, 147	*Dicyphus pallicornis*	**284**
Crow	118, 204, **388-389**	*Dicyphus stachydis*	**284**

Index

Dingy Skipper	**132**
Dinocampus coccinellae	**211**
Diplolepis rosae	**266-267**
Dipper	**323**, 372
Diprion pini	**288**
Ditropis pteridis	**203**
Dog's Mercury	**43**
Donacia sp.	**275**
Dot Moth	**208**
Double-striped Pug	**80**
Dragonflies	44, 111, 136, **140-141**, 156-157, **218-219**, 262, 269, **285**, 300
Black-tailed Skimmer	**220**
Darter, Black	**300**
Darter, Common	**220**
Darter, Ruddy	**220**
Four-spotted Chaser	**140-141**
Hawker, Brown	**218-219**
Hawker, Common	**285**
Hawker, Hairy	**141**
Hawker, Migrant	**285**
Drymus sylvaticus	**278-279**
Duckweed	58, 232-233
Dunnock	**379**
Dyscritulus sp.	**328**
Eagle	368
Ear Moth	**260**
Early Bumblebee	**52-53**, 129
Early Grey	**56-57**
Early Marsh-orchid	**184**
Early Mining Bee	**84-85**
Early Purple Orchid	78, **184**
Earwig, Common	**382-383**
Lesne's Earwig	382
Lesser Earwig	382
Ectemnius lapidarius	**224**
Ectoedemia argyropeza	**366-367**
Ectoedemia subbimaculella	366
Elephant Hawk-moth	**209**
Elfin Saddle	**333**
Elm	69, 248, 265, **306-307**
Emerald Damselfly	**218**
Emperor Moth	**118-119**, **122-123**
Enchanter's-nightshade	**282**
Entomophthora fungus	**87**
Epistrophe eligans	**110**
Ergot	**244**
Eriocampa ovata	**173**, 308
Eriophyes similis	**181**
Eristalis horticola	**210**
Eristalis intricaria	**252-253**, **314-315**
Eristalis pertinax	**109**
Eupeodes corollae	**283**
Eupeodes luniger	**110**
Eutomostethus ephippium	**308-309**
Euura ferruginea	**280**
Euura papillosa	**288**
Euura pavida	**138-139**, 288, 340
Evacanthus interruptus	**265**
Eyebright	**223**
Eyed Hawk-moth	**209**
Eyed Ladybird	**41**
Eyelash Fungus	**167**
Fairy Flax	**304-305**
Fairy Inkcap	**320**
Fallow Deer	**354-355**
False Oxlip	**78**
Feathered Thorn	**362**
Feathery Bog Moss	**338-339**
Feehan, John, *Wildflowers of Offaly*	101
Fenella nigrita	**147**
Fenusella nana	**308-309**
Fern	20, 194, 250, 339
Maidenhair Spleenwort	**20-21**
Polypody	**20-21**

Step into Nature

Fern, Royal Fern	194	Fox Moth	**164-165**
Rustyback	**20-21**	Frochan	100
Wall-rue	**20-21**	Frog	24, **36**, 40, 89, 205, 322, 363, 368, 386
Field Bird's Nest	348	Frog-bit	232-233
Field Forget-me-not	**125**	Froghopper, Common	221
Field Gentian	**270-271**	**Fungi,** *Agaricus campestris*	326
Field Grasshopper	210, **314-315**	Beech Jellydisc	369
Field Madder	79	Birch Bolete	**342-343**
Field Scabious	99, **222**	Blue Roundhead	**350-351**
Field Wood-rush	93	Bog Beacon	88
Fig Gall	**306-307**	Brain, Crystal	369
Figure of Eight	**336**	Brain, Yellow	**18-19**
Figwort Weevil	174	Candlesnuff Fungus	**18-19**
Fine-leaved Sandwort	**158-159**	Collared Earthstar	333
Finnamore Lakes	119, 124, 143, 164, 172-173, 233, 245, 255, 266, 268, 274-275, 278, 281, 297-298, 302, 366	Coral Spot	358
		Coral, Upright	**358-359**
		Dark Honey Fungus	348
Flower Crab Spider	**126-127**, 252-253	Deceiver	**350-351**
Fly Agaric	**326**	Deceiver, Amethyst	**364-365**
Fly	69, 74, **87**, **91-92**, **103**, 107, **127**, 146, **235**, 241, **252-253**, **256**, 257, 258, 295, 297, **306-307**, 319, 326, 328, 334, **377**, 378, 387	*Entomophthora*	**87**
		Ergot	244
		Eyelash	167
Knapweed Gall-fly	253	Fibrecap, Lilac	364
Midge, Owl	**91-91**	Field Bird's Nest	348
Noon Fly	127	Fly Agaric	326
Sieve-winged Snail Killer	**387**	Funnel, Clouded	**350-351**
Twin-lobed Deerfly	**234-235**	Funnel, Tawny	**350-351**
Urophora quadrifasciata	253	Green Elfcup	**320-321**
Yellow-bodied Black Fungus Gnat	243	Holly Parachute	377
		Holly Speckle	377
Forest Shieldbug (Red-legged)	234	Inkcap, Fairy	320
Forget-me-not	125	Inkcap, Glistening	**18-19**, 320
Formica sp.	**380**	Jellybaby	376
Four Coloured (Forest) Cuckoo Bee	129	Lawyer's Wig (Shaggy Ink Cap)	**327**
Four-banded Longhorn Beetle	**216-217**	Lemon Disco	358
Four-spotted Chaser	**140-141**	Oysterling	343
Four-spotted Orbweb Spider	**298-299**	Penny Bun	**342-343**
Fox	**24-25**, 145		
Foxglove	284		

Index

Fungi, Puffball, Common **318**, 333, 348
Puffball, Pestle **348-349**
Purple Jellydisc **358**
Saddle, Elfin **333**
Saddle, White **333**
Scarlet Caterpillarclub **19-20**
Shaggy Ink Cap (Lawyer's Wig) **327**
Shaggy Parasol **321**
Stinkhorn **318-319**
Sulphur Tuft **320-321**, **350-351**
Waxcap, Parrot **350-351**, **364**
Witch's Butter **18-19**
Wood Blewit **364**
Wrinkled Club **369**
Yellow Stagshorn **326**
Gall, aphid, Fig Gall **306-307**
  Pemphigus bursarius **266-267**
Gall, fly, *Chirosia grossicauda* **307**
  Kiefferia pericarpiicola **256**
Gall, fungi, Alder Tongue **266-267**
  Taphrina pruni **181**
Gall, midge, *Dasineura ulmaria* **149**
  Hartigiola annulipes **306**
  Iteomyia major **266-267**
  Jaapiella veronicae **149**
Gall, mite, *Acalitus longisetosus* **149**
  Eriophyes similis **181**
Gall, wasp, *Andricus curvator* **355**
  Andricus kollari **355**
  Andricus quercuscalicis **355**
  Diplolepis rosae **266-267**
  Neuroterus sp. **316-317**
  Pea Gall **316**
  Robin's Pincushion **266-267**
  Silk Button Gall **317**
  Spangle Gall, Common **317**
  Spangle Gall, Smooth **316-317**

Galls   48, **149**, 181, 240, 253, 256, 266-267, 304, **306-307**, **316-317**, 355, 377
Garden Bumblebee **52-53**, 224
Garden Chafer **217**
Garden pond 36, 58-59, 323, 386
Garden Spider **298-299**
Garden Tiger **120**
Garlic Mustard **94**
Germander Speedwell **149**
Ghost Moth **176-177**
Glass Snail **386**
Glistening Inkcap **18-19**, 320
Gloster **124**
Glyphotaelius pellucidus **190-191**
Goat Willow **49**
Goat's-beard **174-175**
Gold Spot **260-261**
Golden Plover **374**
Goldfinch **379**
Good Friday Grass **93**
Gorse   18, 29, **30**, 32, 80, 83, 85, 98, 110-112, 121, 133, 141, 150-151, 203, 285, 375
Gorse Shieldbug **30-32**
Gorse Weevil **32**
Goulson, Dave, *A Sting in the Tail* 52
Grass Emerald **121**
Grasshopper 201, **274-275**, 300, **314-315**
  Common Grasshopper **301**
  Common Green **274-275**, 201
Grasshopper, Field 210, **314-315**
  Large Marsh **300**
  Mottled **201**
Grass-of-Parnassus **286**, 409
Great Pond Snail **386-387**
Great Sundew **101-102**
Great Tit **378**
Great Willowherb **226-227**, 288, 310

418

Greater Butterfly Orchid	185	Heather Ladybird	**112-113**
Greater Stitchwort	**69**, 86, 124	Hebrew Character	**56-57**
Green Elfcup	**320-321**	Hedge Woundwort	231, 284
Green Hairstreak	133	Hedgehog	**116-117**
Green Shieldbug	24, **341**	*Heliophanus sp.*	**298-299**
Green Tiger Beetle	**64-65**	*Helophilus pendulus*	**109**
Green-brindled Crescent	**331**	*Hemichroa crocea*	**188-189**
Greenfinch	9, **378-379**	Hemp-agrimony	240
Green-veined White	43, **96**, 241	Hemp-agrimony Plume	240
Green-winged Orchid	184	*Hepiopelmus variegatorius*	**276**
Grey Arches	208	Herb Bennet (Wood Avens)	134
Grey Heron	**363**	Herb-robert	39
Grey Scalloped Bar	**260**	*Heterarthrus nemoratus*	**188-189**
Grey Shoulder-knot	**336-337**	*Heterocordylus tibialis*	**203**
Grey Wagtail	**323**	*Heterotoma planicornis*	**152-153**
Groundhopper, Slender	**301**, 201	Hieroglyphic Ladybird	**278**
Ground-ivy	42	Hogweed	102, 224, 258
Grypocoris stysi	**170**	Holly	80, 98, **377-378**
Guelder Rose	332	Holly Blue	**98-99**
Gypsywort	**258**	Holly Leaf Gall Fly	**377**, 378
Hagenella *clathrata*	**192**	Holly Parachute	**377**
Hairy Hawker	**141**	Holly Speckle	**377**
Hairy Shieldbug	**32**, **341**	Honeysuckle	55, 185, 248
Halesus radiatus	**328**	Hooded Crow	**389**
Harding, J. M., *The Irish Butterfly Book*	70, 98	House Martin	**145**
Hare	118	House Sparrow	**18**, 379
Harpocera thoracica	**154**	**Hoverfly**	46, 54, 95, **107-111**, 112, 155, 171, 182-183, 210, 224-225, 252-255, 283, 296, 314-315
Hart's-tongue	**250**		
Hartigiola annulipes	**306**		
Haworth's Minor	**260**	**Hoverfly,** *Chrysotoxum bicinctum*	**182**
Hawthorn	29, 70, 85, **130-131**, 202, 281, 325, 331, **332**, 336, 341	*Ectemnius lapidarius*	224
		Epistrophe eligans	**110**
Hawthorn Shieldbug	**341**	*Eristalis horticola*	**210**
Hazel	18, **34-35**, 48, 69, 116, 202, 279, 284, 296, 324	*Eristalis intricaria*	**252-253**, **314-315**
		Eristalis pertinax	**109**
Heather	60, 80, 112, 118, 119, 165, 207, 260, 278, 289, 313, 339, 373, 374	*Eupeodes corollae*	**283**
		Eupeodes luniger	**110**
Heather Beetle	**278**	*Helophilus pendulus*	**109**

Index

Hoverfly, *Leucozona glaucia* **225**
   Leucozona laternaria **225**
   Leucozona lucorum **155**, **255**
   Marmalade **171**
   Platycheirus granditarsus **296**
   Ramsons **107**
   Rhingia campestris **108-109**
   Scaeva pyrastri **183**
   Sericomyia silentis **254-255**
   Sphaerophoria scripta **254**
   Syritta pipiens **110**
   Volucella bombylans **210**, **225**
   Volucella pellucens **225**
   Xylota segnis **210**
Humblebee **50**
Humming-bird Hawk-moth **330**
Ichneumon 68, **252-253**, **276-277**, **340**, 360-316
   Campodorus sp. **340**
   Ichneumon sarcitorius **360-361**
   Mesochorus sp. **340**
Irish Stoat **204**
Ivy Bee 289
Ivy-leaved Toadflax **39**
Jaapiella veronicae **149**
Jackdaw **389**
Jay **389**
Jellybaby **376**
Johns, Rev., *Botanical Rambles* 334
Juniper Shieldbug **44**
Kestrel 368
Kiefferia pericarpiicola **256**
Killaun Bog 24-25, 31, 33, 56, 60-63, 66, 88-89, 100-101, 117, 119-123, 127, 136-137, 161, 178, 188, 193-195, 203, 206-209, 238, 243-244, 260, 273, 285, 288, 292-293, 307, 314, 345, 347, 354, 377
Killyon 213, 266
Kingstown, Birr 166, 256, 288, 367

Kinnitty Demesne 109, 115, 284, 297, 302, 321, 343, 352-353, 358-359, 364-365, 369, 370
Knapweed Gall-fly 253
Knapweed, Common 6-7, 9, 199, **252-253**, 254, 310
Knockbarron 77, 109, 282, 319, 383
Knotted Pearlwort **258-259**
Lacehopper, Spotted **155**
Ladybird, 10-spot **381**
Ladybird, 14-spot **41**, **211**, **278**, **381**
Ladybird, 22-spot **381**
Ladybird, 2-spot **278**
Ladybird, 7-spot **24**, **41**
Ladybird, Cream-spot **113**
Ladybird, Eyed **41**
Ladybird, Heather **112-113**
Ladybird, Hieroglyphic **278**
Ladybird, Orange **113**
Ladybird, Pine **112-113**
Ladybirds **24**, **41**, **44**, **112-113**, **211**, **278**, **381**
Lady's Smock **94-97**, **150**, **155**
Lapwing **374**
Large Carder Bee 52
Large Emerald **207**
Large Marsh Grasshopper **300**
Large Red Damselfly **111**
Large White Butterfly **199**, **277**
Large-flowered Evening-primrose **294-295**
Large-grooved Diving Beetle **59**
Lasius sp. **380**
Lawson's Cypress **44**
Lawyer's Wig (Shaggy Ink Cap) **327**
Leafcutter Bee **84**, **264**, **272-273**
Leafhopper **265**, **279**
Leafmine, fly, Holly Leaf Gall Fly **377**, 378
   Phytomyza chaerophylli **103**
   Phytomyza ilicis **377**, 378

Leafmine, moth, *Acrolepia autumnitella* **344**
   China-mark, Brown   233
   Coleophora deauratella   290-291
   Ectoedemia argyropeza   366-367
   Ectoedemia subbimaculella   366
   Hemp-agrimony Plume   240
   Leucoptera laburnella   **344**
   Leucoptera lotella   **344**
   Parornix devoniella   324
   Phyllonorycter coryli   324
   Phyllonorycter nigrescentella   **344**
   Phyllonorycter oxyacanthae   **325**
   Phyllonorycter spinicolella   **325**
   Psychoides filicivora   250
   Stigmella betulicola   366
   Stigmella hemargyrella   324
   Stigmella luteella   366
   Stigmella oxyacanthella   **325**
   Stigmella plagicolella   **325**
   Stigmella tityrella   324
Leafmine, sawfly, *Fenella nigrita* **147**
   Fenusella nana   **308-309**
   Heterarthrus nemoratus   **188-189**
Lemon Disco   **358**
Leptopterna dolabrata   202
Lesne's Earwig   382
Lesser Butterfly-orchid   **185**
Lesser Celandine   **26-27**, 74, 106
Lesser Diving Beetle   **58-59**
Lesser Earwig   382
Lesser Thorn-tipped Longhorn Beetle **170**
Lesser Water-plantain   **196-197**
Leucoptera laburnella   **344**
Leucoptera lotella   **344**
Leucozona glaucia   225
Leucozona laternaria   225
Leucozona lucorum   **155**, 255
Leuctra fusca   **370-372**

Lichen   20, 29, **33**, 65, 81, 208, 331, 336
Lichen, Cup   **33**
Lichen, Devil's Matchstick   **33**
Lichen, Reindeer   65
Light Emerald   **206**
Lilac Beauty   **207**
Lilac Fibrecap   **364**
Lime tree   172, 248, 284
Limnephilus auricula   **190**
Limnephilus lunatus   **190**
Limnephilus sp.   **37**
Lindman, C., *Bilder ur Nordens Flora*   270
Liocoris tripustulatus   **152-153**
Location, Ballinagar   117
   Ballykealy   124
   Ballylin Bog   303, 331, 344
   Banagher   70, 374
   Birr Demesne   26, 43, 75, 117, 154, 172, 179, 196, 205, 210, 212, 254, 333, 344-346
   Burren   159, 179, 190, 204, 208, 243, 272, 287, 298
   Callows   184
   Charleville Demesne   104, 106, 315, 369, 383
   Clara Bog, Esker and oardwalk 69-70, 72, 100-101, 165, 168, 185, 265, 280, 301
   Cloghan Lakes   112, 196, 231, 251
   Clonfinlough   146, 286
   Clonmacnoise   94, 143, 148, 196, 339, 386
   Clooneen Bog   101, 113, 139, 143, 154, 160, 169, 189, 201, 220, 252, 260, 285, 298, 342
   Cooleeny   76, 78-79
   Cranberry Bog, Gallen, (Noggusboy Bog) 31-32, 62, 65, 110-115, 120-121, 124, 132, 139, 141, 143, 150, 170, 190, 193, 203, 205, 209, 214-215, 218-219, 241, 243, 245, 251, 256, 264, 266, 271, 278, 280-281, 289, 292, 301, 304, 333, 341, 344, 384
   Crinkill   44, 147

Index

Location, Daingean 115
Derryounce Lake 310-311, 313
Finnamore Lakes 119, 124, 143, 164, 172-173, 233, 245, 255, 266, 268, 274-275, 278, 281, 297-298, 302, 366
Gloster 124
Killaun Bog 24-25, 31, 33, 56, 60-63, 66, 88-89, 100-101, 117, 119-123, 127, 136-137, 161, 178, 188, 193-195, 203, 206-209, 238, 243-244, 260, 273, 285, 288, 292-293, 307, 314, 345, 347, 354, 377
Killyon 213, 266
Kingstown, Birr 166, 256, 288, 367
Kinnitty Demesne 109, 115, 284, 297, 302, 321, 343, 352-353, 358-359, 364-365, 369, 370
Knockbarron 77, 109, 282, 319, 383
Lough Boora 33, 140, 185, 218, 251, 259, 326, 374-375
Lusmagh bird hide 374
Mongan Bog 300, 339
Mount St Joseph 78, 146
Pallas Lake 112, 189, 191, 193, 196-197, 203, 218, 241, 294
Ridge Road, Birr 130, **158-159**, 174-175, 217, 287
Shannonbridge 232
Tullamore 28, 77, 147, 251, 298
Turraun Bog 30, 32, 99, 112-113, 156-157, 160-161, 164, 167, 171, 176, 179, 182, 192, 196, 199, 216-220, 235, 237-239, 257, 285

Long-tailed Tit 28-29
Lords-and-Ladies **90-92**
Lough Boora 33, 140, 185, 218, 251, 259, 326, 374-375
Lousewort **134**, 222
Lucerne Plant Bug **302**
Lunar Underwing **336-337**
Lusmagh bird hide 374
Lustrous Bog Moss **338-339**

Lygus wagneri **302**
Magpie (bird) 389
Magpie (moth) 198
Maidenhair Spleenwort **20-21**
Malachite Beetle **150-151**
Malacocoris chlorizans **279**
Mallard **374-375**
Malthinus flaveolus **163**
Malthodes marginatus **163**
Mammal 24-25, 76, 116-118, 145, **204-205, 354-355**, 368

    Bank Vole 76
    Deer, Fallow **354-355**
    Fox **24-25**, 145
    Hare 118
    Hedgehog **116-117**
    Otter 205
    Pine Marten 204
    Rabbit **117-118**, 204, 368
    Red Squirrel **116-118**
    Stoat, Irish 204

Marbled White Spot 179
March Moth **56-57**
Marmalade Hoverfly 171
Marsh Cinquefoil 194
Marsh Damselbug **345**
Marsh Fritillary 167
Marsh Helleborine 231
Marsh Marigold 115
Martin, Sand 145
Mayfly **156**, 323, 371, 372
Cloeon dipterum 156
Meadow Brown 199
Meadow Thistle 194
Meadowsweet 148-149, 256
Megachile sp. 84, 264, **272-273**
Megachile willughbiella **272-273**

Mellinus arvensis	**297**	**Moth,** Burnished Brass	**206**
Mesochorus sp.	**340**	Character, Hebrew	**56-57**
Metatrichia floriformis	**369**	China-mark, Beautiful	**232-233**
Metatropis rufescens	**282**	China-mark, Brown	**233**
Mexican Fleabane	**23**	Chinese Character	**177**
Micropterix calthella	**115**	Clouded Drab	**198**
Migrant Hawker	**285**	*Cochylimorpha straminea*	**252-253**
Millipede, Pill	**383**	*Coleophora deauratella*	**290-291**
Mirid bug	**152-154, 170, 202-203, 226-227, 256-257, 258, 279, 296, 302-303, 314-315**	December Moth	**390-391**
		Diamond-back	**55, 178**
Mompha subbistrigella	**373**	Dot Moth	**208**
Mongan Bog	**300, 339**	Double-striped Pug	**80**
Moorhen	**375**	Ear Moth	**260**
Moss	**20, 24, 29, 39, 50, 60, 88, 100, 116, 118, 151, 165, 263, 279, 331, 338-339, 352, 364, 376**	Early Grey	**56-57**
		Emerald, Grass	**121**
		Emerald, Large	**207**
Broom Fork Moss	**338-339**	Emerald, Light	**206**
Feathery Bog Moss	**338-339**	Emperor	**118-119, 122-123**
Lustrous Bog Moss	**338-339**	Figure of Eight	**336**
Papillose Bog Moss	**338-339**	Moth	**164-165**
Pointed Spear-moss	**338-339**	Fritillary, Marsh	**167**
Red Bog Moss	**338-339**	Garden Tiger	**120**
Sphagnum	**88, 100, 102, 338-339**	Ghost Moth	**176-177**
Moth Trap	**55-57, 66, 190, 292, 312, 390**	Gold Spot	**260-261**
Moth	**20, 40, 46, 48-49, 54-57, 66-69, 80-81, 91, 115, 118-123, 132, 134, 142-143, 161, 164-166, 171, 176-179, 185, 198, 206-209, 214-215, 230, 232, 233, 236-240, 248-250, 253, 260-261, 290-293, 295, 312-313, 324-325, 330**	Green-brindled Crescent	**331**
		Grey Arches	**208**
		Grey Scalloped Bar	**260**
		Hawk-moth, Elephant	**209**
		Hawk-moth, Humming-bird	**330**
		Hawk-moth, Narrow-bordered Bee	**143**
Moth, Angle Shades	**178**	Hawk-moth, Poplar	**176-177**
Argyresthia brockeela	**230**	Haworth's Minor	**260**
Argyresthia goedartella	**230**	Hemp-agrimony Plume	**240**
Bagworm	**166-167**	Hook-tip, Pebble	**120**
Black Rustic	**313**	Hook-tip, Scalloped	**120**
Bordered Beauty	**236**	Lilac Beauty	**207**
Brick	**373**	Lunar Underwing	**336-337**
Burnet Companion	**143**		

Index

Moth, Magpie 198
Marbled White Spot 179
March Moth 56-57
Micropterix calthella 115
Mompha subbstrigella 373
Mother of Pearl 248-249
Muslin Footman 208
Nemophora minimella 237
Nettle-tap 178
Oak Beauty 81
Oak Eggar 121
Pebble Prominent 121
Peppered Moth 206
Plutella porrectella 54-55
Plutella xylostella 55, 178
Psyche casta 166-167
Psychoides filicivora 250
Purple and Gold, Common 143
Purple Clay 237
Purple-barred, Small 143
Puss Moth 66-68, 214-215
Quaker, Common 56-57
Quaker, Red-line 336-337
Rusty-dot Pearl 312
Sallow 292-293
Shark 248-249
Shoulder Stripe 81
Shoulder-knot, Blair's 336-337
Shoulder-knot, Grey 336-337
Silver Hook 179
Six-spot Burnet 179
Small Chocolate-tip 238-239
Spectacle 208
Sprawler, The 361
Swallow-tailed Moth 206
Thorn, Canary-shouldered 248-249
Thorn, Feathered 362
Tiger, Ruby 209

Moth, Tussock, Dark 119
Tussock, Pale 119
Vapourer 313
Vestal, The 312
White Plume 279
Winter Moth 390
Ypsolopha mucronella 80-81
Mother of Pearl 248-249
MothsIreland 56
Mottled Grasshopper 201
Mount St Joseph 78, 146
Mountain Everlasting (Cat's Paw) 146
Muslin Footman 208
Myopa sp. 127
Myrmica sp. 380
Mystacides azurea 190
Narrow-bordered Bee Hawk-moth 143
Nasturtium 132, 199
National Biodiversity Data Centre 9, 56
Nematinus steini 173
Nemophora minimella 237
Neriene montana 385
Nettle 22-23, 71-72, 89, 152, 155, 170, 178, 248, 256, 258, 278, 302-303, 361, 385
Nettle-tap 178
Nomada goodeniana 85-86
Nomada marshamella 85-86
Noon Fly 127
Nursery Web Spider 89
Oak 81, 154, 202, 234, 283, 316-317, 321, 322, 355, 366
Oak Beauty 81
Oak Eggar 121
Oblong-leaved Sundew 102
Oecetis ochracea 191
Opposite-leaved Golden-saxifrage 39
Orange Ladybird 113

Step into Nature

Orange-legged Furrow Bee	**84**, 128	*Phytomyza ilicis*	**377**
Orange-tip	**94-97**	Pied Wagtail	**377**, 378
Orchid	**78-79, 175, 184-185**, 199, **287**	Pignut	**102-103**
Orchid, Butterfly-, Greater	185	Pill Millipede	**383**
Orchid, Butterfly-, Lesser	**185**	*Pinalitus cervinus*	**284**
Orchid, Early Marsh-orchid	**184**	Pine	24, 115, 288
Orchid, Green-winged	**184**	Pine Ladybird	**112-113**
Orchid, Pyramidal	**185**, 199	Pine Marten	**204**
Otter	**205**	Pineappleweed	**79**
Owl Midge	**91-92**	*Pithanus maerkelii*	**279**
Oxeye Daisy	54, 99, 244, 302	*Plagiognathus arbustorum*	**303**
Oysterling	**343**	*Plagiognathus chrysanthemi*	**303**
Pachygnatha clerckii	**385**	**Plant,** Alexanders	**43**
Painted Lady	**200**	Avens, Water	**134-135**
Pale Tussock	**119**	Avens, Wood (Herb Bennet)	134
Pallas Lake	112, 189, 191, 193, 196-197, 203, 218, 241, 294	Bilberry	**100**
Pantilius tunicatus	**296**	Bird's-foot-trefoil	128, 132, **134**, 143, 179, 244, 275, 310, 344
Papillose Bog Moss	**338-339**	Bladderwort	**294-295**
Parent Shieldbug	**160-161, 246-247**	Bluebell	70, 74, **104-106**, 108
Paronix devoniella	**324**	Bog Asphodel	**194-195**
Parrot Waxcap	**350-351, 364**	Bog Rosemary	**100**
Pea Gall	**316**	Bogbean	**124**
Peacock	47, **71**, 72, 241	Bracken	188-189, 203, 307
Pebble Hook-tip	**120**	Brambles	188, 199, 256, 263
Pebble Prominent	**121**	Broom	80, 118, 121, 203
Pemphigus bursarius	**266-267**	Burnet Rose	**159**
Penny Bun	**342-343**	Buttercup	27, 115
Peppered Moth	**206**	Butterwort, Common	**124-125**, 295
Perforate St John's-wort	**304**	Cat's Paw (Mountain Everlasting)	146
Pestle Puffball	**348-349**	Centaury, Common	222, 310
Phillips, Roger, *Mushrooms*	351	Chickweed, Common	**22-23**, 45, 124
Phryganea sp.	**192**	Cinquefoil, Creeping	**147**
Phyllonorycter coryli	**324**	Cinquefoil, Marsh	194
Phyllonorycter nigrescentella	**344**	Clover	128, 244, **290-291**, 302, 310-311, 313
Phyllonorycter oxyacanthae	**325**	Clover, Red	**290-291**, 310-311
Phyllonorycter spinicolella	**325**	Coltsfoot	**26-27**, 42
Phytomyza chaerophylli	**103**		

Index

Plant, Cow Parsley 94, 102-**103**
   Cowslip **77-78**
   Cow-wheat **222-223**
   Cranesbill, Bloody Crane's-bill **159**
   Cuckooflower **94-97**, 150, 155
   Daisy 27, 79, **334-335**
   Daisy, Oxeye 54, 99, 244, 302
   Dame's Violet **54-55**, 178
   Dandelion 27, 50, 71, 85-86, 95, 129, 175
   Dog's Mercury **43**
   Enchanter's-nightshade **282**
   Eyebright 223
   Fairy Flax **304-305**
   False Oxlip **78**
   Field Forget-me-not **125**
   Field Madder **79**
   Field Wood-rush **93**
   Figwort, Common **174**
   Fine-leaved Sandwort **158-159**
   Fleabane, Blue **251**, 310
   Fleabane, Common **294-295**
   Fleabane, Mexican **23**
   Forget-me-not **125**
   Garlic Mustard **94**
   Gentian, Autumn **270-271**
   Gentian, Field **270-271**
   Goat's-beard **174-175**
   Good Friday Grass 93
   Gorse 18, 29, **30**, 32, 80, 83, 85, 98, 110-112, 121, 133, 141, 150-151, 203, 285, 375
   Grass-of-Parnassus **286**
   Ground-ivy **42**
   Gypsywort **258**
   Hart's-tongue **250**
   Heather 60, 80, 112, 118, 119, 165, 207, 260, 278, 289, 313, 339, 373, 374
   Hedge Woundwort **231**, 284
   Hemp-agrimony **240**

Plant, Herb Bennet (Wood Avens) **134**
   Herb-robert **39**
   Hogweed **102**, 224, 258
   Honeysuckle 55, 185, 248
   Ivy 42, 80, 98, 242, 284, 313, **332**, 337, 373, 385
   Knapweed, Common 6-7, 9, 199, **252-253**, 254, 310
   Knotted Pearlwort **258-259**
   Lady's Smock **94-97**, 150, 155
   Large-flowered Evening-primrose **294-295**
   Lesser Celandine **26-27**, 74, 106
   Lesser Water-plantain **196-197**
   Lords-and-Ladies **90-92**
   Lousewort **134**, 222
   Marjoram, Wild 71, **241**
   Marsh Helleborine **231**
   Marsh Marigold **115**
   Meadow Thistle **194**
   Meadowsweet **148-149**, 256
   Mountain Everlasting (Cat's Paw) **146**
   Mouse-ear, Common **45**
   Nasturtium 132, 199
   Nettle **22-23**, 71-72, 89, 152, 155, 170, 178, 248, 256, 258, 278, 302-303, 361, 385
   Nettle, Red Dead- **22-23**
   Opposite-leaved Golden-saxifrage **39**
   Orchid **78-79**, **175**, **184-185**, 199, 287
   Orchid, Autumn Lady's-tresses **287**
   Orchid, Bee **175**
   Orchid, Common Twayblade **79**
   Orchid, Early Purple **78**, **184**
   Pignut **102-103**
   Pimpernel, Bog **196-197**
   Pimpernel, Yellow **134**
   Pineappleweed **79**
   Primrose **77-78**
   Purple Moor-grass 244, 260
   Pimpernel, Bog **196-197**

Step into Nature

Plant, Pimpernel, Yellow	134	**Plant,** Violet	74, 262-263
Pineappleweed	79	Water Mint	71, **241**, 310
Primrose	77-78	Whitlowgrass, Common	**45**
Purple Moor-grass	244, 260	Wild Angelica	258
Purple-loosestrife	**228-229**	Wild Garlic (Ramsons)	**106-107, 135**
Quaking-grass	**159**	Willowherb, Great	**226-227**, 288, 310
Ramsons (Wild Garlic)	**106-107, 135**	Winter Aconite	**26-27**
Red Bartsia	**231**	Wood Anemone	**74-77**
Sanicle	**146**	Woodruff	**125**
Scabious, Devil's-bit	167, 173, 237, 254-255, 297	Wood-sorrel	**76**
Scabious, Field	99, **222**	Yarrow	303, 310
Shepherd's-purse	**22-23**	Yellow Corydalis	**147**
Skullcap	**196-197**	Yellow-rattle	**148**
Snowdrop	**26-27**	Yellow-wort	**241**
Soapwort	**270-271**	*Platycampus luridiventris*	**308-309**
Speedwell, Common Field-	**42**	*Platycheirus granditarsus*	**296**
Speedwell, Germander	149	*Plutella porrectella*	**54-55**
Spindle	70, 81, 198, **332**	Pocket Plum Gall	181
St John's-wort, Perforate	**304**	Pointed Spear-moss	**338-339**
Stitchwort, Bog	**124**	Polypody	**20-21**
Stitchwort, Greater	69, 86, 124	Pond Skater, *Gerris* sp.	62
Strawberry, Barren	38	Pondweed	232-233
Strawberry, Wild	38	Poplar (Tree)	66, 138, 266
Sundew	**101-102**, 124, 157, 295	Poplar Hawk-moth	**176-177**
Thistles	129, 169, 194, 199, 200, 248, 254, 256, 263, 278, 315	Potato Capsid Bug	303
Toadflax, Ivy-leaved	**39**	Pratt, Anne	8
Toadflax, Purple	**357**	Primrose	**77-78**
Toothwort	**69**	*Pristiphora conjugata*	172
Tormentil	**147**	*Pristiphora leucopus*	172
Traveller's-joy	**294-295**	*Protonemura meyeri*	372
Trifid bur-marigold	**270**	*Psyche casta*	**166-167**
Valerian, Common	**196-197**	*Psychoides filicivora*	**250**
Valerian, Red	196, 248	Purple Clay	**237**
Vetch	**125**, 128-129, 302, 315, **332**, 344	Purple Jellydisc	**358**
Vetch, Bush	**125, 332**, 344	Purple Moor-grass	244, 260
Vetch, Common	**125**	Purple Toadflax	**357**
		Purple-loosestrife	**228-229**

Index

Puss Moth	**66-68, 214-215**	Sand Martin	145
Pyramidal Orchid	**185**, 199	Sanicle	146

Quaker, Common — **56-57**
Quaking-grass — 159

Rabbit — **117-118**, 204, 368
Raft Spider — **89**
Ragged Robin — 196
Ramsons (Wild Garlic) — **106-107**, 135
Ramsons Hoverfly — **107**
Raspberry Sawfly — **309**
Raven — **388**
Red Admiral — 72, **99**
Red Bartsia — **231**
Red Bog Moss — **338-339**
Red Clover — **290-291, 310-311**
Red Dead-nettle — **22-23**
Red Squirrel — **116-118**
Red Valerian — 196, 248
Red-headed Cardinal Beetle — **150-151**
Red-legged (Forest) Shieldbug — **234**
Red-line Quaker — **336-337**
Red-tailed Bumblebee — 9, **52-53**, 210, 222
Rhingia campestris — **108-109**
Ridge Road, Birr — 130, **158-159**, 174-175, 217, 287
Ringlet — 199
Robin's Pincushion — **266-267**
Rook — **388-389**
Round-leaved Sundew — **101-102**
Rowan — 116, 292, 325
Royal Fern — **194**
Ruby Tiger — **209**
Ruby-tailed Wasp — **162-163**
Ruddy Darter — **220**
Rustyback — **20-21**
Rusty-dot Pearl — **312**
Sallow — **292-293**
Sand Martin — 145

Sawfly — 48, **138-139**, **147**, **155**, 161, 171, **172-173**, **186-189**, 257, **280-281**, **288-289**, **308-309**, 340

Abia nitens — 173
Aglaostigma aucupariae — **155**
Allantus sp. — **188-189**
Arge ustulata — **281**
Caliroa cerasi — **257**
Cimbex connatus — **186-187**
Cimbex femoratus — **139**, 186
Cimbex luteus — **186**
Cladius brullei — **309**
Cladius grandis — **288**
Diprion pini — **288**
Eriocampa ovata — 173, 308
Eutomostethus ephippium — **308-309**
Euura ferruginea — **280**
Euura papillosa — **288**
Euura pavida — **138-139, 288, 340**
Fenella nigrita — **147**
Fenusella nana — **308-309**
Hemichroa crocea — **188-189**
Heterarthrus nemoratus — **188-189**
Nematinus steini — **173**
Platycampus luridiventris — **308-309**
Pristiphora conjugata — **172**
Pristiphora leucopus — **172**
Rasberry Sawfly — **309**
Strongylogaster multifasciata — **188**
Tenthredo colon — **288**
Scaeva pyrastri — **183**
Scale Insect — **163**
Scalloped Hook-tip — **120**
Scarce Blue-tailed Damselfly — **268-269**
Scarlet Caterpillarclub — **19-20**

Scots Pine	24, 288	**Slime Mould**, *Trichia varia*	352
Sericomyia silentis	**254-255**	Small Blue	98
Shaggy Ink Cap (Lawyer's Wig)	327	Small Chocolate-tip	**238-239**
Shaggy Parasol	321	Small Copper	133
Shannonbridge	232	Small Heath	**180-181**
Shark	**248-249**	Small Purple-barred	143
Sharp-tail Bee	**264**, 273	Small Scabious Mining Bee	297
Sheildbug, Juniper	44	Small Tortoiseshell	**71-72**
Sheldrake, Merlin, *Entangled Life*	20	Small White	101, **132**, 199
Shepherd's-purse	**22-23**	Smooth Newt	**60-62**
Shieldbug	24, 30-31, 32, 44, 160-161, 168, 234, 246-247, 257, 341	Smooth Spangle Gall	**316-317**
		Snail, Amber	386

Shieldbug:
- Birch 257, **341**
- Blue **168**
- Bronze **161, 341**
- Forest (Red-legged) 234
- Gorse **30-32**
- Green 24, **341**
- Hairy **32, 341**
- Hawthorn **341**
- Parent **160-161, 246-247**
- Red-legged (Forest) 234
- Spiked **161,** 234
- Tortoise Bug **257**

Snail:
- Brown Lipped Snail 386
- Garden Snail 386
- Glass Snail 386
- Great Pond Snail **386-387**
- Strawberry Snail 386

Shoulder Stripe	81	Snapdragon	357
Sialis sp.	**193**	Snipe	356
Sicus ferrugineus	127, **252-253**	Snowdrop	**26-27**
Sieve-winged Snail Killer	387	Soapwort	**270-271**
Silk Button Gall	317	Sparrowhawk	368
Silver Hook	179	Speckled Wood	**346-347**
Silver-washed Fritillary	**262-263**	Spectacle	208
Siskin	82	*Sphaeridium* sp.	**328-329**
Six-spot Burnet	179	*Sphaerophoria scripta*	254
Skullcap	**196-197**	Sphagnum moss	88, 100, 102, **338-339**
Slender Groundhopper	301, 201	**Spider**	29, 36, 82, **89, 126-127, 252-253,** 257, **283, 298-299,** 345, 372, 385
Slime mould	248, **352-353,** 369		

Slime mould:
- *Ceratiomyxa fruticulosa* **352-353**
- *Metatrichia floriformis* 369

Spider:
- *Arctosa perita* **298-299**
- Candy-striped Spider 283
- Flower Crab Spider **126-127, 252-253**
- Four-spotted Orbweb **298-299**
- Garden Spider **298-299**
- *Heliophanus* sp. **298-299**
- *Neriene montana* 385
- Nursery Web Spider 89
- *Pachygnatha clercki* 385

Index

Spider, Raft	**89**
Sputnik Spider	**283**
Zebra Jumping Spider	**298-299**
Zygiella x-notata	**385**
Spiked Shieldbug	**161**, 234
Spindle	70, 81, 198, **332**
Spittlebug	221
Spotted Lacehopper	**155**
Sprawler, The	**361**
Sputnik Spider	**283**
Starling	**40**, 292
Stenodema laevigata	**152-153**
Stenophylax permistus	**192**
Step, Edward, *Nature Rambles* **8-9**, 50
Stigmella betulicola	**366**
Stigmella hemargyrella	**324**
Stigmella luteella	**366**
Stigmella oxyacanthella	**325**
Stigmella plagicolella	**325**
Stigmella tityrella	**324**
Stinkhorn	**318-319**
Stonechat	**375**
Stonefly	193, **370-373**
Isoperla grammatica	**370**
Leuctra fusca	**370-372**
Protonemura meyeri	**372**
Strawberry Snail	**386**
Strongylogaster multifasciata	**188**
Sulphur Tuft	**320-321, 350-351**
Sundew	**101-102**, 124, 157, 295
Great Sundew	**101-102**
Oblong-leaved	102
Round-leaved	**101-102**
Swallow	**144-145**
Swallow-tailed Moth	**206**
Swift	**144-145**
Syritta pipiens	**110**
Tanyptera atrata	**136**

Taphrina pruni	**181**
Tawny Funnel	**350-351**
Teasel	54
Tenthredo colon	**288**
Tetrix sp.	**301**
Thistles	129, 169, 194, 199, 200, 248, 254, 256, 263, 278, 315
Thread-legged Bug	**345**
Toothwort	**69**
Tormentil	**147**
Tortoise Bug	**257**
Traveller's-joy	**294-295**
Tree, Alder	82, 138, 154, 173, 186-187, 189, 208, 230, 234, 248, 266-267, 296, 308-309
Apple	177, 331, **332**, 336
Aspen	66, 172, 214-215, 288, 366-367
Birch	41, 82, 87, 112, 115-116, 118, 120, 136-137, 139, 149, 154, 161, 163, 173, 182, 189, 206, 217, 230, 237, 242, 246-248, 278, 280-281, 283, 292, 296, 308-309, 313, 326, 341, 366
Blackthorn	**46-48**, 83, 85, 177, 181, 257, 313, 325, 331, **332**, 336
Buckthorn	70, 98, 163, 190-191
Crab Apple	331, **332**, 336
Elm	69, 248, 265, 306-307
Hawthorn	29, 70, 85, **130-131**, 202, 281, 325, 331, **332**, 336, 341
Hazel	18, **34-35**, 48, 69, 116, 202, 279, 284, 296, 324
Holly	80, 98, 377-378
Lime	172, 248, 284
Oak	81, 154, 202, 234, 283, 316-317, 321, 322, 355, 366
Pine	24, 115, 288
Poplar	66, 138, 266
Rowan	116, 292, 325
Scots Pine	24, 288
Willow	46, **48-49**, 66, 71, 83, 85, 128, 138, 176-177, 209, 236, 238-239, 266-267, 280-281, 288, 292, 237, 340

Treecreeper	242	White Saddle	333
Trichia varia	352	White-tailed Bumblebee	**49-50**, 210
Trifid Bur-marigold	270	Whooper Swan	30
Tullamore	28, 77, 147, 251, 298	Wild Angelica	258, 281
Turraun Bog	30, 32, 99, 112-113, 156-157, 160-161, 164, 167, 171, 176, 179, 182, 192, 196, 199, 216-220, 235, 237-239, 257, 285	Wild Carrot	256
		Wild Garlic (Ramsons)	**106-107**, 135
		Wild Marjoram	71, **241**
Twin-lobed Deerfly	**234-235**	Wild Strawberry	38
Two-banded Longhorn Beetle	**114-115**	Willow	46, **48-49**, 66, 71, 83, 85, 128, 138, 176-177, 209, 236, 238-239, 266, 267, 280-281, 288, 292, 237, 340
Upright Coral	358-359		
Urophora quadrifasciata	253	Willowherb	152, 209, **226-227**, 288, 310, 373
Vapourer	313	Winter Aconite	26-27
Variable Damselfly	142	Winter Moth	390
Vestal, The	312	Witch's Butter	**18-19**
Vetch	125, 128-129, 302, 315, **332**, 344	Wood Anemone	**74-77**
Violets and Common Dog-	74, 262-263	Wood Avens (Herb Bennet)	134
Volucella bombylans	**210**, 225	Wood Blewit	364
Volucella pellucens	225	Wood White	132
Wall-rue	**20-21**	Woodruff	125
Ward, Mary, *Sketches with the Microscope*	331, 383	Wood-sorrel	76
		Wren	379
Wasp	68, 103, 107, 127, **162-163**, 171, 174, 182, 186, **211**, **224-225**, **252-253**, 254, 266-267, **276-277**, **297**, **316-317**, 328, **340**, 355, **360-361**	Wrinkled Club	369
		Xylota segnis	210
		Yarrow	303, 310
		Yellow Brain	**18-19**
Ancistrocerus gazella	**162-163**	Yellow Corydalis	147
Ancistrocerus trifasciatus	**171**	Yellow Pimpernel	134
Chalcid	253	Yellow Stagshorn	326
Cotesia glomeratus	**277**	Yellow Water-lily	232-233
Dinocampus coccinellae	**211**	Yellowhammer	82
Dyscritulus sp.	**328**	Yellow-rattle	**148**
Ichneumon	68, **252-253**, **276-277**, **340**, 360-316	Yellow-wort	**241**
Mellinus arvensis	**297**	*Ypsolopha mucronella*	**80-81**
Ruby-tailed	**162-163**	**Zebra** Jumping Spider	**298-299**
Water Avens	**134-135**	*Zonocyba bifasciata*	**265**
Water Cricket	**58**	*Zygiella x-notata*	**385**
Water Measurer	**62**		
Water Mint	71, **241**, 310		
White Emrine	276		
White Plume Moth	**279**